Hands-on Deep Learning

A Guide to Deep Learning
with Projects and Applications

Harsh Bhasin

Apress®

Hands-on Deep Learning: A Guide to Deep Learning with Projects and Applications

Harsh Bhasin
Faridabad, Haryana, India

ISBN-13 (pbk): 979-8-8688-1034-3   ISBN-13 (electronic): 979-8-8688-1035-0
https://doi.org/10.1007/979-8-8688-1035-0

Copyright © 2024 by Harsh Bhasin

This work is subject to copyright. All rights are reserved by the Publisher, whether the whole or part of the material is concerned, specifically the rights of translation, reprinting, reuse of illustrations, recitation, broadcasting, reproduction on microfilms or in any other physical way, and transmission or information storage and retrieval, electronic adaptation, computer software, or by similar or dissimilar methodology now known or hereafter developed.

Trademarked names, logos, and images may appear in this book. Rather than use a trademark symbol with every occurrence of a trademarked name, logo, or image we use the names, logos, and images only in an editorial fashion and to the benefit of the trademark owner, with no intention of infringement of the trademark.

The use in this publication of trade names, trademarks, service marks, and similar terms, even if they are not identified as such, is not to be taken as an expression of opinion as to whether or not they are subject to proprietary rights.

While the advice and information in this book are believed to be true and accurate at the date of publication, neither the authors nor the editors nor the publisher can accept any legal responsibility for any errors or omissions that may be made. The publisher makes no warranty, express or implied, with respect to the material contained herein.

Managing Director, Apress Media LLC: Welmoed Spahr
Acquisitions Editor: Celestin Suresh John
Development Editor: Laura Berendson
Coordinating Editor: Gryffin Winkler

Cover designed by eStudioCalamar

Cover image designed by Freepik (www.freepik.com)

Distributed to the book trade worldwide by Apress Media, LLC, 1 New York Plaza, New York, NY 10004, U.S.A. Phone 1-800-SPRINGER, fax (201) 348-4505, e-mail orders-ny@springer-sbm.com, or visit www.springeronline.com. Apress Media, LLC is a California LLC and the sole member (owner) is Springer Science + Business Media Finance Inc (SSBM Finance Inc). SSBM Finance Inc is a **Delaware** corporation.

For information on translations, please e-mail booktranslations@springernature.com; for reprint, paperback, or audio rights, please e-mail bookpermissions@springernature.com.

Apress titles may be purchased in bulk for academic, corporate, or promotional use. eBook versions and licenses are also available for most titles. For more information, reference our Print and eBook Bulk Sales web page at http://www.apress.com/bulk-sales.

Any source code or other supplementary material referenced by the author in this book is available to readers on GitHub (https://github.com/Apress). For more detailed information, please visit https://www.apress.com/gp/services/source-code.

If disposing of this product, please recycle the paper

To My Mother ...

Table of Contents

About the Author .... **xiii**

About the Technical Reviewers .... **xv**

Acknowledgments .... **xix**

Chapter 1: Revisiting Machine Learning **1**

    Machine Learning: Brief History, Definition, and Applications................................... 3

    Types of Machine Learning: Task (T) 6

    Performance (P) 7

    Conventional Machine Learning Pipeline 11

    Regression 12

    Feature Selection 14

        Filter Method 14

        Wrapper Method 18

        Filter vs. Wrapper Methods... 19

    Feature Extraction 19

        Gray-Level Co-occurrence Matrix... 20

        Local Binary Pattern 21

        Histogram of Oriented Gradients 24

    Principal Component Analysis... 24

    Bias–Variance Trade-off 28

        Overfitting and Underfitting 28

        Bias and Variance 29

    Application: Classification of Handwritten Digits Using a Conventional Machine Learning Pipeline 30

TABLE OF CONTENTS

    Conclusion 38

    Exercises 39

        Multiple-Choice Questions 39

        Applications 41

    References 42

Chapter 2: Introduction to Deep Learning 43

    Neurons 43

    From Perceptron to the Winter of Artificial Intelligence 45

    Imagery and Convolutional Neural Networks 47

    What's New 49

    Sequences 50

    The Definition 51

    Generate Data Using Deep Learning 52

    Conclusion 55

    Exercises 56

        Multiple-Choice Questions 56

        Activity 57

    References 58

Chapter 3: Neural Networks 59

    Objectives 59

    Introduction 59

    Single-Layer Perceptron 62

    Implementation of a SLP 64

    XOR Problem 75

    Activation Functions 76

        1. Sigmoid 76

        2. Tanh 77

        3. Rectified Linear Unit (ReLU) 78

        4. Softmax 79

TABLE OF CONTENTS

- Multi-layer Perceptron 80
  - Solving the XOR Problem Using Multi-layer Perceptron 80
  - Architecture of MLP and Forward Pass 82
- Gradient Descent 84
- Backpropagation 86
- Implementation 87
- Conclusion 104
- Exercises 105
  - Multiple-Choice Questions 105
  - Theory 108
  - Numerical 109
- References 109

Chapter 4: Training Deep Networks 111

- Introduction 111
- Train–Test Split 111
- Train–Validation–Test Split 112
- K-Fold Split 112
- Batch, Stochastic, and Mini-batch Gradient Descent 113
  - Batch Gradient Descent 114
  - Stochastic Gradient Descent 114
  - Mini-batch Gradient Descent 114
- RMSprop 116
- Adam Optimizer 118
- Conclusion 126
- Exercises 127
  - Multiple-Choice Questions 127
  - Theory 129
  - Experiments 130
- References 130

vii

Chapter 5: Hyperparameter Tuning 133
Introduction 133
Bias–Variance Revisited 134
Hyperparameter Tuning 137
Experiments: Hyperparameter Tuning 142
Conclusion 150
Exercises 150
    Multiple-Choice Questions 150
    Experiments 154
References 155

Chapter 6: Convolutional Neural Networks: I 157
Convolutional Layer 159
Implementing Convolution 161
Padding 165
Stride and Other Layers 167
    Stride 167
    Pooling 168
    Normalization 169
    Fully Connected Layer 170
Importance of Kernels 170
Architecture of LeNet 177
Conclusion 180
Exercises 182
    Multiple-Choice Questions 182
    Numerical 184
    Applications 184

Chapter 7: Convolutional Neural Network: II 185
Sequential Model 186
    Creating the Model 186
    Adding Layers in the Model 187

TABLE OF CONTENTS

Removing the Last Layer from the Model 187
Initializing Weights 188
Summary 188
Keras Layers 189
1. Dense Layer 189
2. Conv2D Layer 190
3. Pooling 190
4. Activations 190
5. Initializing Weights 191
6. Miscellaneous 191
MNIST Dataset Classification Using LeNet: Prerequisite 192
LeNet 192
Structure 192
Implementation 194
AlexNet 198
Some More Architectures 201
GoogLeNet 201
ResNet 201
DenseNet 202
Conclusion 202
Exercises 202
Multiple-Choice Questions 202
Implementations 205
References 205

Chapter 8: Transfer Learning 207
Introduction 207
Idea 207
VGG 16 and VGG 19 for Binary Classification 208
Types and Strategies 217

TABLE OF CONTENTS

    Limitations and Applications of Transfer Learning 219

    Conclusion 220

    Exercises 220

        Multiple-Choice Questions 220

        Application 222

    References 222

Chapter 9: Recurrent Neural Network 225

    Introduction 225

    Why Neural Networks Cannot Infer Sequences 226

    Idea 228

    Backpropagation Through Time 229

    Types of RNN 230

    Applications 234

        Sentiment Classification 234

        Parts of Speech Tagging 241

        Handwritten Text Recognition 249

        Speech to Text 250

    Conclusion 251

    Exercises 251

        Multiple-Choice Questions 251

        Theory 254

        Image Captioning 254

    References 255

Chapter 10: Gated Recurrent Unit and Long Short-Term Memory 257

    Introduction 257

    GRU 258

    Long Short-Term Memory 260

    Named Entity Recognition 262

    Sentiment Classification 273

    Conclusion 282

Exercises	283
Multiple-Choice Questions	283
Theory	285
Application-Based Questions	285
References	286

Chapter 11: Autoencoders 287

Introduction	287
Concept and Types	287
The Math	288
Types of Autoencoders	288
Autoencoder and Principal Component Analysis	290
Training of an Autoencoder	291
Latent Representation Using Autoencoders	293
Experiment 1	293
Experiment 2	297
Finding Latent Representation Using Multiple Layers	299
Variants of Autoencoders	302
Sparse Autoencoder	302
Denoising Autoencoder	303
Variational Autoencoder	303
Conclusion	303
Exercises	304
Multiple-Choice Questions	304
Theory	306
Applications	306

Chapter 12: Introduction to Generative Models 307

Introduction	307
Hopfield Networks	307
Boltzmann Machines	310
A Gentle Introduction to Transformers	314

TABLE OF CONTENTS

    An Introduction to Self-Attention 315
        The Transformer 317
  Conclusion 318
  Exercise 318
        Multiple-Choice Questions 318
        Theory 320
  References 321

Appendix A: Classifying The Simpsons Characters 323

Appendix B: Face Detection 331

Appendix C: Sentiment Classification Revisited 335

Appendix D: Predicting Next Word 343

Appendix E: COVID Classification 347

Appendix F: Alzheimer's Classification 351

Appendix G: Music Genre Classification Using MFCC and Convolutional Neural Network 355

Index 359

About the Author

Harsh Bhasin is a researcher and practitioner. He has completed his PhD in "Diagnosis and Conversion Prediction of Mild Cognitive Impairment Using Machine Learning" from Jawaharlal Nehru University, New Delhi. He worked as a Deep Learning consultant for various firms and taught at various universities, including Jamia Hamdard and Delhi Technological University (DTU). He is currently associated with Bennett University.

He has authored 11 books including *Programming in C#*, Oxford University Press, 2014, and *Algorithms*, Oxford University Press, 2015. He has authored more than 40 papers that have been published in international conferences and renowned journals, including *Alzheimer's & Dementia*, *Soft Computing*, Springer *Nature*, *BMC Medical Informatics and Decision Making*, *AI & Society*, etc. He is the reviewer of a few renowned journals and has been the editor of a few special issues. He has been a recipient of Visvesvaraya Fellowship, Ministry of Electronics and Information Technology.

His areas of expertise include Deep Learning, algorithms, and medical imaging. Apart from his professional endeavors, he is deeply interested in Hindi poetry: the progressive era and Hindustani classical music: percussion instruments.

About the Technical Reviewers

Karanbir Singh is an accomplished engineering leader with over 7 years of experience leading AI/ML engineering, distributed systems, and microservices projects across diverse industries, including fintech and automotive. Currently working as a Senior Software Engineer at Salesforce, he focuses on backend technologies as well as AI. His career has been marked by a commitment to building high-performing teams, driving technological innovation, and delivering impactful solutions that enhance business outcomes.

At TrueML, as an engineering manager, he managed a critical team to develop and deploy Machine Learning models in production. He successfully expanded and led engineering teams, significantly improving feature development velocity and client engagement through strategic collaboration and mentorship. His leadership directly contributed to increased revenue, client retention, and substantial cost savings through innovative internal solutions. His role involved not only steering technical projects but also shaping the company's roadmap in partnership with data science, product management, and platform teams.

Previously, at Lucid Motors and Poynt, he developed critical components and integrations that advanced product capabilities and strengthened industry partnerships. His technical expertise spans across AI/ML, cloud computing, and software architecture, and he is adept at utilizing cutting-edge technologies and methodologies to drive results.

Karanbir holds a master's degree in Computer Software Engineering from San Jose State University and has been recognized for his innovative contributions, including winning the Silicon Valley Innovation Challenge. He is passionate about mentoring and coaching emerging talent and thrives in environments where he can leverage his skills to solve complex problems and advance technological initiatives.

ABOUT THE TECHNICAL REVIEWERS

Prashanth Josyula, a dynamic force in the tech world whose journey is marked by an unyielding passion for innovation and an extraordinary depth of expertise in both technical literature and software engineering. As a Principal Member of Technical Staff (PMTS) at Salesforce, Prashanth doesn't just meet expectations—he consistently exceeds them, pushing the boundaries of what's possible in technology.

With over 16 years of robust experience in the IT industry, Prashanth has mastered a multitude of programming languages and technologies, establishing himself as a true polyglot programmer. His proficiency spans across Java, Python, Scala, Kotlin, JavaScript, TypeScript, Shell Scripting, SQL, and an array of open source solutions. Since beginning his professional journey in 2008, he has delved into various domains, each time leaving a mark of excellence.

In the realm of **Java/Java EE and Spring**, Prashanth has been instrumental in designing and building resilient, scalable backend systems that power critical applications across industries. His deep understanding of these technologies ensures robust and high-performance solutions tailored to meet complex business needs.

Prashanth's expertise in **UI technologies** is equally impressive. He has crafted intuitive, responsive user interfaces using frameworks like ExtJS, JQuery, DOJO, Angular, and React. His commitment to creating seamless user experiences shines through in every project, bridging the gap between complex backend processes and user-friendly frontend interfaces.

Venturing into **big data**, Prashanth has leveraged platforms like Hadoop, Spark, Hive, Oozie, and Pig to transform massive datasets into valuable insights, driving strategic decisions and innovations. His ability to harness the power of big data showcases his analytical mindset and his knack for tackling large-scale data challenges.

In the field of **microservices and infrastructure**, Prashanth has been a pioneer, in engineering robust and scalable solutions with cutting-edge tools like Kubernetes, Helm, Terraform, and Spinnaker. His contributions to open source projects reflect his commitment to collaborative innovation and continuous improvement.

Moreover, Prashanth is at the forefront of **AI and Machine Learning**, exploring and advancing the capabilities of these transformative technologies. His work in this area is characterized by a fearless approach to experimentation and a relentless pursuit of knowledge.

Each day for Prashanth is an exciting adventure, filled with opportunities to learn, innovate, and lead. His career is a testament to his dedication to advancing technology, not just for the sake of progress, but to truly make a difference. With his unparalleled skills and a visionary mindset, Prashanth continues to inspire peers and push the envelope of technological possibility.

Acknowledgments

Knowledge is in the end based on acknowledgement.

—*Ludwig Wittgenstein*

I have been lucky enough to have met people who inspired me to learn. First of all, I would like to thank Professor Moinuddin, former Pro-Vice Chancellor, Delhi Technological University, for his unconditional support. He has deposed his faith in me in my formative years and helped me grow. I would also like to thank the late Professor A. K. Sharma, former Dean and Chairperson, Department of Computer Science, YMCA, Faridabad, for his constant encouragement. I have been able to write this book, author papers, and work on projects only because of the encouragement provided by him. I am also thankful to the following academicians and professionals for their encouragement and providing unconditional support to me:

- Professor I. K. Bhat, Vice Chancellor, MRU, India
- Professor Prashant Jha, King's College London
- Professor Tapas Kumar, Associate Dean, SET, MRIIRS, India
- Professor Ranjit Biswas, former Dean, Faculty of Engineering, Jamia Hamdard
- Professor Naresh Chauhan, Department of Computer Science, YMCA University of Science and Technology

I am thankful to my student Nishant Kumar, NorthCap University, for editing the chapters. I would also like to acknowledge the help of the students' team, Amit Thakur, Ankit Singh, and Jai Mishra, for their help.

I would like to express my sincere gratitude to my mother, Vanita Bhasin; sister, Swati Bhasin; and rest of the family, including my pets, late Zoe and Xena, and friends for their unconditional support to me.

ACKNOWLEDGMENTS

I am extremely grateful to the team at Springer for their insightful guidance and unwavering support. I am also thankful to the team and reviewers for their editorial feedback, design inputs, and constant reviews, which have transformed this manuscript into an informative and interesting book.

I would be glad to receive your comments or suggestions, which can be incorporated in the future editions of the book. You can reach me at `i_harsh_bhasin@yahoo.com`.

—Dr. Harsh Bhasin

CHAPTER 1

Revisiting Machine Learning

Imagine being transported back to the late 1990s in the United States, where the authorities discover your expertise in Machine Learning (ML). They reach out to seek your assistance in the automation of a time-consuming task: reading pin code on letters. Supposedly there are 500 such employees in various post offices across the country, and each employee was being paid a sum of $2000 per month to perform this task. This accumulates to a monthly expenditure of $1000000, resulting in an annual cost of $12 million, or a staggering $60 million nationwide, over the next five years.

To assist the government in saving valuable exchequer funds, you are tasked with designing a program that can efficiently read and interpret pin codes on the letters. This solution will not only help in cost savings but also will greatly augment accuracy and accelerate the process. Can you think of an algorithm to accomplish this task?

It turns out that it is not very easy to write such an algorithm. Let's see why! To understand the problem, let us start with an algorithm that recognizes "1" in a 28-pixel × 28-pixel image. Ideally, the pixels around the central vertical may be considered for identifying if the image contains "1." However, the number is handwritten, and therefore it can be written in many ways, in terms of scale, orientation, style, etc. Figure 1-1 shows some pictures of handwritten "1"s obtained from the popular MNIST dataset containing images of handwritten digits. If recognizing "1" is difficult, then imagine recognizing all the digits and alphabets and processing these, in general.

CHAPTER 1   REVISITING MACHINE LEARNING

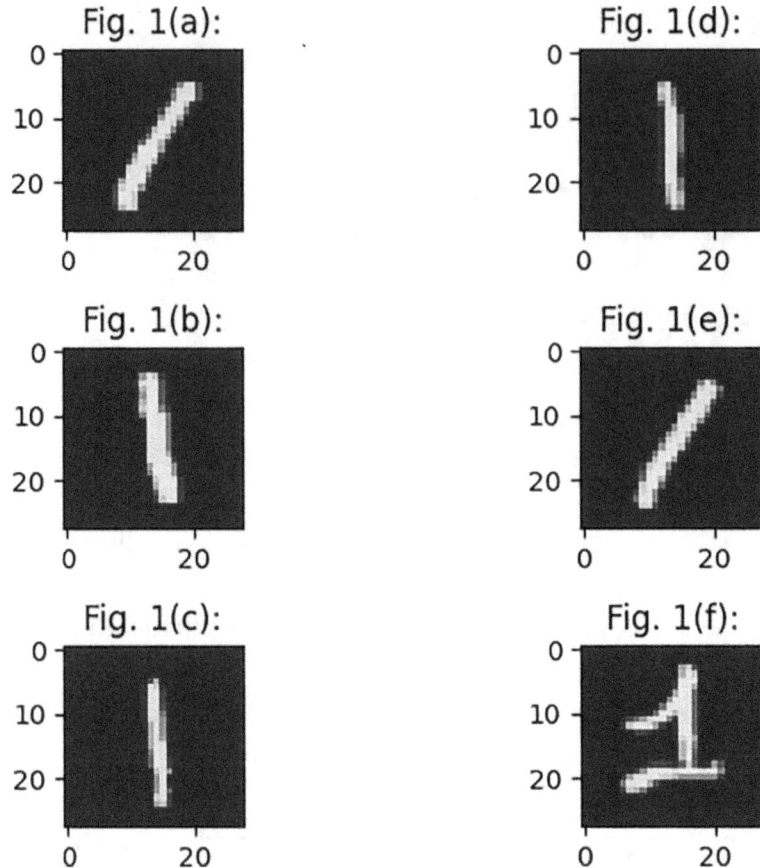

Figure 1-1. *Some pictures of "1"s obtained from the MNIST dataset*

Recognizing handwritten digits is an easy task for human beings, but it is difficult to come out with a set of rules or algorithms that recognizes the digit in a given picture. So we need some system that can imitate human beings to accomplish this task. Here Machine Learning (ML) can help us. Informally ML can be defined as follows:

Machine Learning *is a subset of Artificial Intelligence, which may be considered as the ability of machines to imitate humans.* [1]

The formal definition of ML is discussed in the following sections. ML helps us accomplish tasks like disease classification, prediction and forecasting, object recognition, sentiment classification, etc.

This chapter briefly introduces Machine Learning and discusses its types, the pipeline and its components, its applications, and the bias–variance trade-offs. This chapter also presents **MNIST dataset classification** using a conventional Machine Learning pipeline employing feature extraction, feature selection, classification, and analysis of the results. The chapter also includes the Python implementations of some of the most important feature extraction and selection techniques. Feature extractions from various modalities like images, sound, and text are briefly discussed in this chapter. In addition to the above, the chapter hovers over an important dimension reduction methodology called Principal Component Analysis (PCA). The chapter ends with a case study, namely, the **classification of the MNIST dataset using a conventional Machine Learning pipeline**. The case study uses an important feature extraction technique, Local Binary Pattern (LBP), selects the important features using a filter method, and uses Support Vector Machine (SVM) to classify the data. The reader new to this domain may not be versed with some of the terms used in this section. For such readers, the following sections will be helpful. However, those familiar with these concepts may skip this chapter and move to the next one.

Machine Learning: Brief History, Definition, and Applications

Since time immemorial, humans have been trying to develop machines that are intellectually as good as human beings. The desire of machines to learn as humans do and get better at a task with experience helped us reach the present age of splendid technological advancement. This betterment should be measurable. The development of Checkers Program by Samuel, at IBM, in the 1950s can be considered as one of the initial steps toward this goal. The 1960s saw progress in the field of pattern recognition, particularly after the works of Rosenblatt on perceptron followed by that of Minsky and Papert describing the limitations of perceptron. The 1970s saw the development of expert systems and symbolic natural language processing. The following decade witnessed advancements in Decision Trees and the development of Multi-layer Perceptron (MLP). Some of the most important learning methodologies like Support Vector Machines, Reinforcement Learning, and ensemble models were developed in the 1990s. The desire of the scientific community to develop machines that could beat humans in some cognitive tasks got a boost with the development of Deep Blue, at IBM, which defeated the then-chess champion Garry Kasparov. The work toward designing the self-driven cars, initially using the above methodologies, has come a long way since. Figure 1-2 depicts the major milestones in the journey of Machine Learning till 1999.

CHAPTER 1   REVISITING MACHINE LEARNING

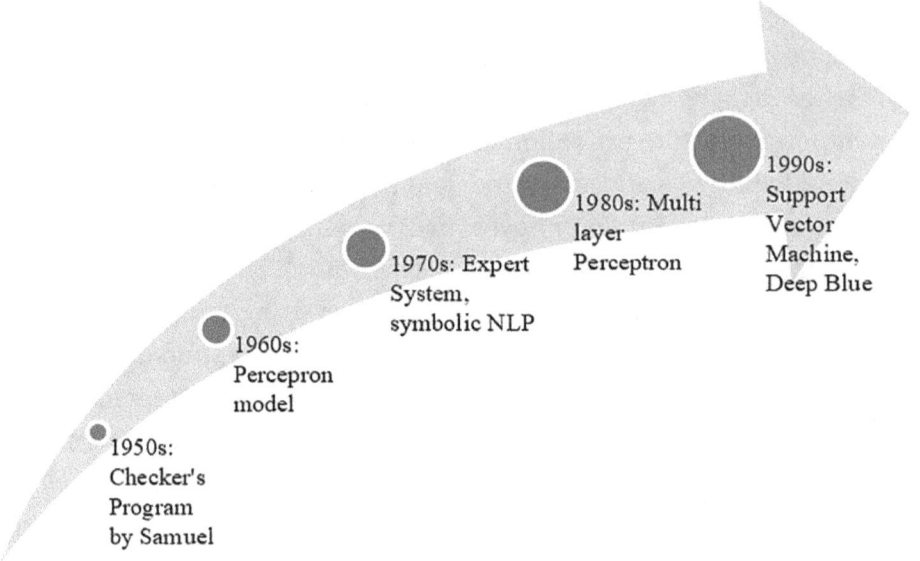

Figure 1-2. *Machine Learning before 2000*

Machine Learning (ML) is a subset of Artificial Intelligence (AI). ML algorithms are trained and tested using datasets and help us do tasks, which humans do better. The dataset may or may not be labeled. AI, on the other hand, strives to develop machines with "human like cognitive abilities" [2]. To understand the concept, let us take an example. Suppose you need to develop a system that takes an image as an input and classifies it as "cat" or "not cat." The input images are of size 100×100, and the output is a binary number having a value 0 (not cat) or 1 (cat). You somehow develop this system and take 1000 new images, out of which the system correctly identifies 673 images. The percentage of unseen images correctly classified (accuracy) is hence 67.3. You ask one of your friends, who happens to be a Machine Learning engineer, to help you improve the system. They modify the system, after which the system correctly classifies 721 images, thus improving the accuracy by 4.8%. Moreover, as the system is trained with more images, the accuracy increases. Considering the percentage of unseen samples correctly identified, that is, accuracy, as the performance measure, the performance, P, of the system improves with experience, E (in this case, data), on the given task, T (classification). This system is therefore learning. Formally, Machine Learning can be defined as

> *A system is said to learn when the performance **P** improves with Experience **E**, on task **T**. [3]*

ML is currently being used in various domains, from product recommendation to stock market prediction, to disease detection, etc. Some of the interesting applications of Machine Learning are as follows:

> Recommendation Systems: Harry had an account on Amazon and started buying his favorite stuff after he received his first salary. He was fond of books, stationary, and music. So he bought books like *The Fault in Our Stars*, fancy notebooks, and a percussion instrument from the platform. He bought similar stuff the next month also. When he visited the platform again, the recommendation section displayed some books by John Green and others, some musical instruments, notebooks, and sound bars. Can you guess why books by John Green and sound bars were shown in the recommendation section? This is because the platform learns using Machine Learning, leveraging user data and ratings. It also uses natural language processing, discussed later in this book. Now visit your YouTube and that of your friend. Just think of the reasons if you find the recommendations for the two different.

> Google Maps: Assume you need to go for an interview to a company located at Gurugram, a city located in the vicinity of New Delhi, the capital of India. You are currently living in Delhi and have never been to that company. You decide to ride a car to reach your destination and find the best route using an app called Google Maps. Wait! How does this app know the best route from your location to the destination? Also, the app claims that some routes are better than others, in terms of congestion, distance, or some other criteria. This app uses Machine Learning to find the optimal path from source to destination. It gets the traffic data from "Waze," an app that Google bought in 2013. If you are using this app since long, you must have observed that its performance has significantly improved. The credit for this also goes to Machine Learning. Well, your turning on location does help Google Maps also.

Other examples of applications of Machine Learning include

- Disease detection and prediction
- Amazon Alexa
- Self-driving vehicles
- Sentiment Analysis
- Customer churning

Each of the above is discussed in detail in the following chapters. Now, you got an idea that Machine Learning is used everywhere: right from the face recognition on your handheld devices to the recommendations in Netflix. Let's move to the types of learning.

Types of Machine Learning: Task (T)

Machine Learning can be classified as supervised, unsupervised, semi-supervised, or reinforcement. In supervised learning, the system is trained using the samples and corresponding labels. During testing, it is given the input, and it generates the predicted output. The learning algorithm tries to learn the parameters of the model to decrease the gap between the predicted label and the correct label. Supervised learning can further be classified as classification and regression. In classification, the labels corresponding to samples are discrete, whereas in the case of regression, they are continuous.

In unsupervised learning, the system is provided with the features, and no label is associated with the samples. These algorithms unveil the patterns in the given data. The examples of such learning include finding trends on social media, the association between the products, etc.

Supervised Learning

"In supervised learning, we are provided with some input/output samples (X, y). The algorithm aims to find a function y = f(X), that relate the feature vector with the label. This function f is learnt and evaluated on some unseen data" [4].

Unsupervised Learning

"In unsupervised learning, we are given only samples X of the data, and we compute a function f such that y = f(X) is *simpler*" [5]. Clustering is a type of unsupervised learning.

Semi-supervised Learning

"Semi-supervised learning (SSL) is halfway between supervised and unsupervised learning. In addition to unlabelled data, the algorithm is provided with the labels of some of the samples, not all" [5].

Reinforcement Learning

"In Reinforcement learning, the system acts on the environment, and it gets some feedback. Based on this feedback the system alters its actions". Reinforcement Learning is often used in automated drones.

The next element in the definition of Machine Learning is performance, P. Let us now understand some of the common performance measures.

Performance (P)

Consider a classification problem having two classes: Positive (P) and Negative (N). To classify this dataset, you design a system which predicts Positive or Negative for an unknown sample. The predictions can be True Positive (TP), True Negative (TN), False Positive (FP), or False Negative (FN). The classification results can be represented in a confusion matrix, as shown in Figure 1-3.

- True Positive (TP): The model correctly predicts a positive instance.

- True Negative (TN): The model correctly predicts a negative instance.

- False Positive (FP): The model incorrectly predicts that an instance is positive, when it is actually negative. This is referred as Type I error.

- False Negative (FN): The model incorrectly predicts that an instance is negative, when it is actually positive. This is referred as Type II error.

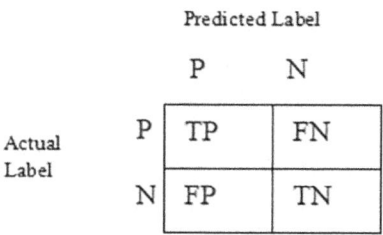

Figure 1-3. *The confusion matrix of a binary classification problem*

These four cases help evaluate the performance of the developed model. Important metrics like accuracy, specificity, recall, precision, and F1 score can be derived from these cases to offer a broad assessment of the model's effectiveness in distinguishing between the two classes. Note that the model should have minimum possible false positives and false negatives, while true positives and true negatives should be as high as possible. Table 1-1 shows the various performance measures for a two-class problem and their brief description.

Table 1-1. Classification Metrics

Performance Measure	Formula	Description	Keras Implementation	sklearn Implementation
Accuracy	$\frac{TP+TN}{TP+TN+FN+FP}$	Total number of test cases correctly classified.	tf.keras.metrics.Accuracy[1]	sklearn.metrics.accuracy_score
Specificity (False Positive Rate)	$\frac{TN}{TN+FP}$	Total number of negative test cases correctly classified_.		
Sensitivity/recall (True Positive Rate)	$\frac{TP}{TP+FN}$	Total number of positive test cases correctly classified.	tf.keras.metrics.Recall1[1]	sklearn.metrics.recall_score
Precision	$\frac{TP}{TP+FP}$	Goodness of positive predictions.	tf.keras.metrics.Precision[1]	precision_score[2]
F-score	(2 × Recall × Precision)/ (Recall + Precision)	It is used for unbalanced class problems, where accuracy may be misleading.	tf.keras.metrics.F1Score[1]	f1_score[3]

CHAPTER 1 REVISITING MACHINE LEARNING

In order to use the functions stated in the table, you need to import the following (refer to the superscript of the functions in the table):

1. import tensorflow as tf
2. sklearn.metrics.precision_score
3. fromsklearn.metricsimport f1_score

For a multiclass problem, the above matrix can be extended as required. For example, for a three-class classification problem, the matrix shown in Figure 1-4 explains the performance of the classifier, not just in terms of correct classifications, but also how many test samples are classified as other classes. The diagonal of this matrix depicts the test cases correctly classified by the algorithm. In *sklearn* it is implemented as *sklearn.metrics.confusion_matrix*.

	Predicted Label		
I	I	II	III
II	I	II	III
II	I	II	III

Actual Label

Figure 1-4. *Confusion matrix*

In the case of a multiclass problem, the class-wise precision and recall can be calculated. The precision and the recall of the model can be perceived as the average precision and average recall of each class. The usage of the above metrics is shown in the examples and illustrations that follow.

The plot of specificity and sensitivity is referred to as the Receiver Operating Curve (ROC) by varying the threshold. The area under this curve is called AUC or Area under the Receiving Curve (Figure 1-5).

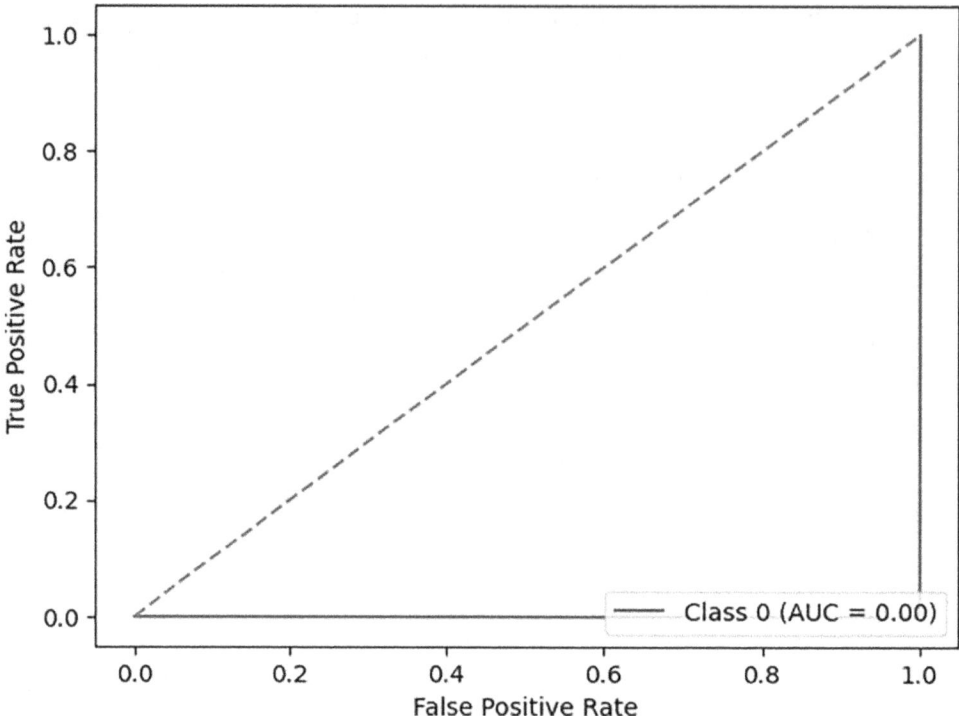

Figure 1-5. *An example of ROC-AUC curve*

The metrics for evaluating the performance of regression are shown in Table 1-2.

Table 1-2. *Regression Metrics*

Performance Measure	Formula	sklearn Implementation	Keras Implementation
Mean Squared Error	$\frac{1}{N}\sum_{i=1}^{N}(y_i - \hat{y})^2$	sklearn.metrics.mean_squared_error	tf.keras.metrics.MeanSquaredError
Root Mean Squared Error	$\sqrt{\frac{1}{N}\sum_{i=1}^{N}(y_i - \hat{y})^2}$	sklearn.metrics.mean_squared_error squared = False, returns RMSE	tf.keras.metrics.RootMeanSquaredError
Mean Absolute Error	$\frac{1}{N}\sum_{i=1}^{N}|y_i - \hat{y}|$	sklearn.metrics.median_absolute_error	tf.keras.metrics.MeanAbsoluteError
R-Squared	$1 - \frac{\frac{1}{N}\sum_{i=1}^{N}(y_i - \hat{y})^2}{\frac{1}{N}\sum_{i=1}^{N}(y_i - \bar{y})^2}$	sklearn.metrics.r2_score	tf.keras.metrics.R2Score

Each of the above is explained in the following chapters, as and when they are used. Now, let us now move to the elements of a conventional Machine Learning pipeline.

Conventional Machine Learning Pipeline

The conventional Machine Learning pipeline includes the complete process of developing a Machine Learning model. This includes steps from data collection to model deployment. The major steps in the Machine Learning pipeline include

> Problem Definition: The problem at hand needs to be clearly defined and classified as a supervised learning, unsupervised learning, or Reinforcement Learning problem.
>
> Data Collection and Preprocessing: The protocol of collecting data is then decided. The data is then collected, and preprocessing including handling missing values, outlier analysis, and other processes aimed at addressing the inconsistencies in the data are carried out.
>
> Exploratory Data Analysis (EDA): This step is essential to analyze the given data and access the characteristics of the data.
>
> Feature Engineering: This step includes selecting relevant features from the existing features, transforming existing features, or creating new features to improve the performance of the model.
>
> Data Splitting: The data is then divided into the train set, validation set, and test set. The train set is used to train the model, the validation set is used to find the values of the hyperparameters, and the model is evaluated using the test set.
>
> Choosing a Model: This is followed by choosing the learning algorithm like Support Vector Machine, Decision Tree, etc.
>
> Model Training: The model is then trained on the training set. The validation set is used to adjust the hyperparameters of the so-formed model. In order to do this, grid search, random search, or other optimization methods are used.

Model Evaluation: The performance of the model is then evaluated using the test set. Metrics uses to do this have already been discussed.

Analysis: The model's decisions are then interpreted based on the application.

Model deployment, monitoring, and maintenance follow. Based on the feedback of the deployed model and the insights, each step may be refined multiple times. Figure 1-6 summarizes the discussion.

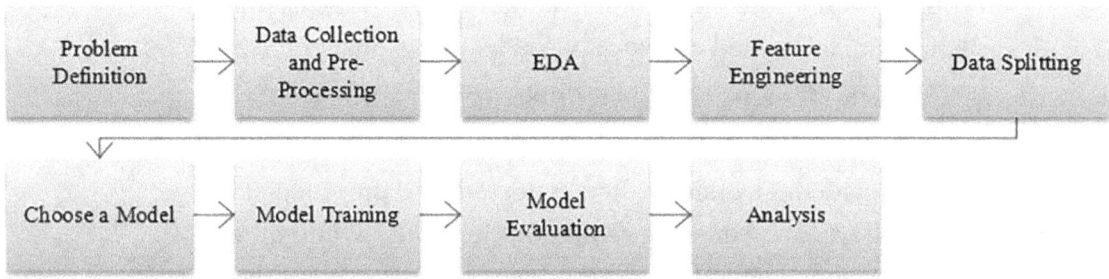

Figure 1-6. *Conventional Machine Learning pipeline*

Let's now have a look at one of the tasks, namely, regression, and understand how we actually learn the parameters of a model in a type of regression called linear regression.

Regression

Regression is a type of supervised learning where we are given (X, y), where $X \epsilon R^d$ and $y \epsilon R$. That is, the labels are continuous. Regression aims to develop a model that predicts y (y_pred) for an unseen X, when the model has been trained on the training data.

The parameters of the model are learned by minimizing the squared difference between y_pred and y_test. That is to minimize $loss = (y_{pred} - y_{test})^2$ or $s = \frac{1}{2}(y_{pred} - y_{test})^2$, where ½ is inserted just for the sake of mathematical convenience.

This loss can be minimized by finding the gradient with respect to the parameter and incrementally moving in the opposite direction.

In the case of linear regression, the label y can be considered as the linear combination of X_m^i for a sample X_m. That is,

$$y\_pred = \sum_{i=1}^{d} w_i X_m^i$$

The values of w_is can be calculated using the concept explained above. That is,

$$loss = \frac{1}{2}(y_{pred} - y_{test})^2$$

- $loss = \frac{1}{2}\left(\sum_{i=1}^{d} w_i X_m^i - y_{test}\right)^2$
- $\partial loss/(\partial w_i) = \partial/(\partial w_i)\left(\frac{1}{2}\left(\sum_{i=1}^{d} w_i X_m^i - y_{test}\right)^2\right)$
- $\partial loss/(\partial w_i) = (y_{pred} - y_{test})X_m^i$
- $-\partial loss/(\partial w_i) = -(y_{pred} - y_{test})X_m^i$

Therefore, after each iteration, the weights are changed as per the following formula:

$$w_i = w_i - \alpha(y_{pred} - y_{test})X_m^i$$

where α is the learning rate.

In general,

$$W = W - \alpha(y_{pred} - y_{test})X_x$$

The value of α determines the step size at each iteration. If the value of this parameter is small, it will take a longer time to reach the optimal solution, whereas if it is large, we may skip the optimal solution. The web resources include the code of linear regression and its application to the popular Boston Housing price dataset.

Note that at times it becomes important to extract features from a given dataset, or reduce the number of features, or transform the features to another space. Feature selection and feature extraction are two of the most important components of a ML pipeline. Let us have a brief overview of both of them.

Feature Selection

Feature selection aims to select a subset of features from among the given features with the aim of minimizing the classification error. That is, for a given $X = \{X^1, X^2, ..., X^n\}$, a subset $X = \{X^1, X^2, ..., X^d\}$, $d \leq n$, of the most representative features is to be selected with the aim of minimizing the memory requirements and the computation time of the model. Feature selection is needed because some of the features do not contribute to enhancing the performance of the model and some may negatively affect the performance of the model.

The readers may note that feature selection is not the same as dimensionality reduction wherein new features may be computed and the original data and units are generally lost. In contrast, in feature selection, only a small amount of features are selected, and original data is preserved. This may also be considered as an optimization problem, wherein a subset of features is selected with the objective of optimizing the objective function.

Feature selection may use search strategies or evaluation strategies. Heuristic search algorithms like genetic algorithms are often used in selecting the optimal subset of features. The evaluation strategies include filter and wrapper methods (Figure 1-7).

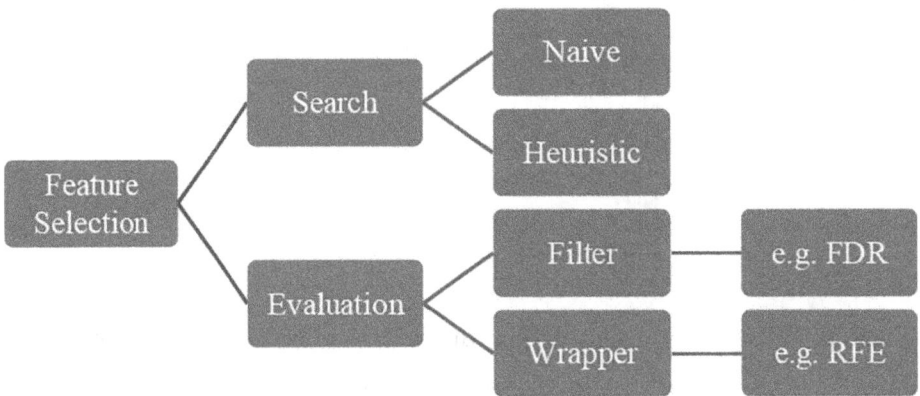

Figure 1-7. Feature selection method

Filter Method

In filter methods the selection of features is independent of the learning algorithm. This may be done with the help of the information content. For example, a feature selection method called Fisher Discriminant Ratio (FDR), generally used for a two-class problem, gives more importance to a feature in which the distance between the centers of the clusters of those two classes is more, whereas the variance of those two clusters is less, that is, for a feature X_i having two subsets X_1 and X_2 representing the data of the two classes.

$$(m_1 - m_2)^2 \text{ is more}$$

whereas

$$(s_1^2 + s_2^2) \text{ is less}$$

where m_1 is the mean of X_1, m_2 is the mean of X_2, s_1 is the standard deviation of X_1, and s_2 is the standard deviation of X_2. The formula for calculating the FDR of a feature is

$$FDR = \frac{(m_1 - m_2)^2}{s_1^2 + s_2^2}$$

This method can be used in Forward Feature Selection (FFS). In FFS, the FDR of each feature is calculated, and the features are ordered in descending order of their FDR values. This is followed by taking the first feature (from the so-ordered dataset) and evaluating the performance in the first iteration. In the second iteration two features are taken and so on. The performance of the model in each iteration is noted, and the minimum number of features that result in optimal performance is selected.

The following code shows the arrangement of features in order of their FDR scores for the popular IRIS dataset, followed by the application of Forward Feature Selection.

Code:

```
#Importing Libraries
from sklearn.datasets import load_iris
import numpy as np
from sklearn.model_selection import train_test_split
from sklearn.svm import SVC
from matplotlib import pyplot as plt

#Loading Data
Data= load_iris()
X = Data.data
y = Data.target
```

CHAPTER 1 REVISITING MACHINE LEARNING

```python
X = X[:100, :]
y = y[:100]
print(X.shape, y.shape)

#Calculating FDR
def calFDR(X, y):
    X1 = X[:50,:]
    X2 = X[50:, :]
    m1 = np.mean(X1, axis = 0)
    m2 = np.mean(X2, axis = 0)
    s1 = np.std(X1, axis = 0)
    s2 = np.std(X2, axis = 0)
    fdr = ((m2 - m1)**2)/(s1**2 + s2**2)
    ind = np.argsort(fdr)
    ind= ind[: : -1]
    return fdr, ind

#FDR Output
fdr, ind1= calFDR(X, y)
X1 = X[:,ind1 ]
print(ind1)

#Forward Feature Selection
accuracies = []
for i in range(X.shape[1]):
    X2 = X1[:,:(i+1)]
    X_train, X_test, y_train, y_test = train_test_split(X2, y,
    test_size=0.3)
    clf1 = SVC(kernel='linear')
    clf1.fit(X_train, y_train)
    y_pred = clf1.predict(X_test)
    acc = np.sum(y_pred==y_test)/y_pred.shape[0]
    accuracies.append(acc)
print(accuracies)

#Plotting
X_imp = X[:,2]
```

```
X1 = X_imp[:50]
X2 = X_imp[50:]
ind1 = np.arange(50)
plt.scatter(ind1, X1, label='class 0', color='r')
plt.scatter(ind1, X2, label='class 1', color='b')
plt.title('Scatter Plot')
plt.legend()
plt.show()
```

Note that, in the IRIS dataset, the most important feature (as selected by FDR) can easily classify the two classes as shown in the scatter plot in Figure 1-8.

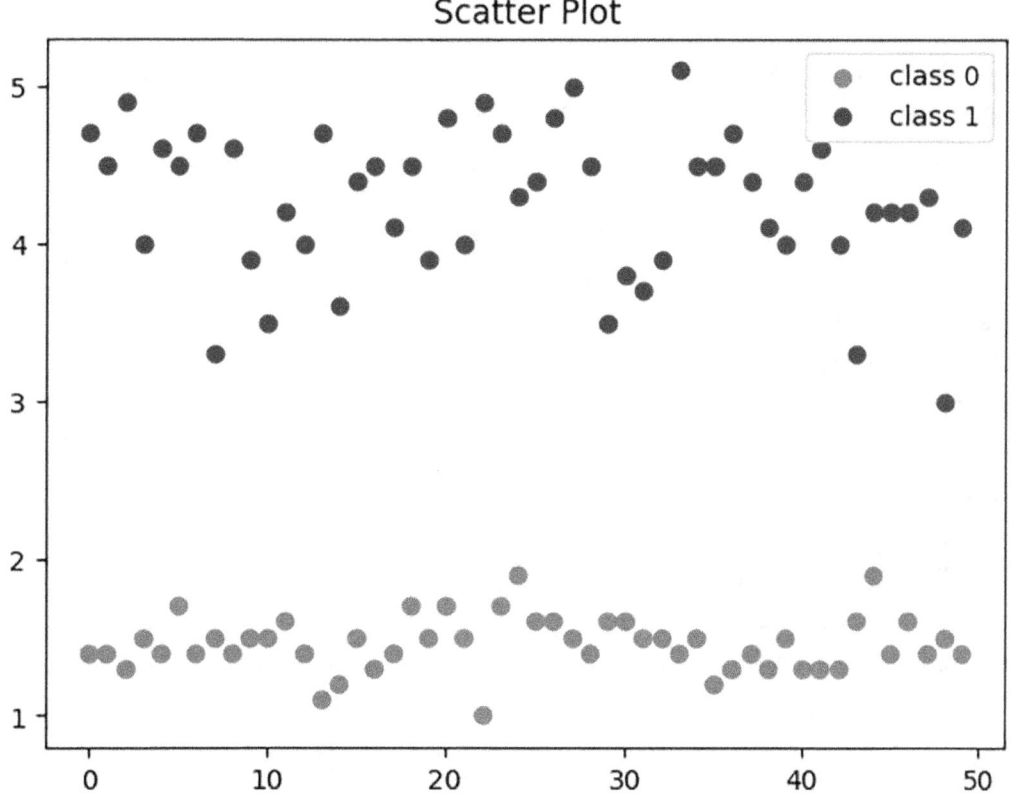

Figure 1-8. *Scatter plot of the two classes in the most important feature selected by FDR*

Wrapper Method

In wrapper methods we order the features in terms of their performance viz-a-viz the classifier. For example, a popular wrapper method called Recursive Feature Elimination (RFE) uses the following strategy:

1. First of all, we take all the features and find the performance with respect to the given classifier.
2. Then we eliminate one feature at a time and note the performances in all the cases.
3. The feature whose removal improves the performance is then eliminated.
4. This process is repeated with the so-obtained subset of features, till no further optimization is possible.

The following code shows the application of RFE on the wine dataset:

```
from sklearn.datasets import load_diabetes
from sklearn.feature_selection import RFE
from sklearn.svm import SVR
#load diabetes dataset
data=load_diabetes()
X=data.data
y=data.target
#Select regression model, in this case SVR
model=SVR(kernel="linear")
#Create feature selector
feat_selector=RFE(model, n_features_to_select=5, step=1)
feat_selector.support_
```

Output:

```
array([False, False,  True,  True, False, False,  True,  True,  True,
       False])
```

feat_selector.ranking_

```
array([4, 6, 1, 1, 3, 5, 1, 1, 1, 2])
```

Filter vs. Wrapper Methods

The filter methods are generally faster and make use of the intrinsic property of the data, though they have a disadvantage: generally they end up selecting a larger subset of features. The wrapper methods, on the other hand, generally lead to better accuracy and avoid overfitting. However, these methods are much slower and are highly sensitive to the selection of a classifier.

Note that feature selection is an exhaustive topic in which there are many more methods like sequential forward selection, sequential backward selection, bidirectional search, and so on. It is difficult to find the most appropriate feature selection method for your task and dataset. Here Deep Learning comes handy as it almost eliminates the need of feature selection.

Feature Extraction

A classification system generally extracts features from the given data before applying classification to it. Feature extraction is needed as more representative, compact representation of the input data is needed to design an effective and efficient system. This section briefly discusses various feature extraction methods particularly used in image analysis. The methods in this section find application in robot vision, medical imaging, character recognition, etc. The feature extraction methods used for text data are discussed in Chapter 3 of this book, and those used for sound data are discussed in the Appendix G. As per *Pattern Recognition* by Theodoridis and Koutroumbas (Elsevier, 2006)

Feature Extraction

"The goal of feature extraction in images is to generate a feature vector which is generally fed into a classifier and helps it to classify images in one of the possible classes."

Feature extraction is not only used in classification but also in segmentation and to reduce redundant information. In addition to the above, raw images generally contain a lot of pixels, and all these pixels cannot be taken as the features of a given image. For example, for a 1024 × 1024 image, the number of pixels is 1 million. If all these pixels are taken as features, then the system will have to learn 1 million parameters that will require a large amount of training data and computation data and a huge amount of memory. If somehow the same image can be represented as a vector containing 256 values, then the system will become much more efficient and effective. As a matter of fact, using the pixels of a raw image as features will lead to the curse of dimensionality. As per Bellman (*Adaptive Control Processes*, Princeton University Press, Princeton, NJ, 1961), the curse of dimensionality can be defined as

Curse of Dimensionality

"The number of samples required to estimate an arbitrary function with the given accuracy grows exponentially with respect to the number of input variables (Dimensionality of the function)."

So reducing the number of features helps us handle the curse of dimensionality. For images many types of features can be extracted. These include

- Histogram features
- Gray-level features
- Shape features
- Color features

Histogram features, also referred to as texture features, generally include either first-order statistics or second-order statistics of the image. The first-order statistics contain information related to the gray-level distribution, whereas the second-order statistics include information related to the relative distribution of gray levels. Examples of second-order gray-level features include co-occurrence matrix.

Gray-Level Co-occurrence Matrix

In Gray-Level Co-occurrence Matrix (GLCM), the co-occurrence matrix of gray levels is calculated. This is followed by evaluation of the direction of orientation with the step size of 45 degree. For each direction we calculate six metrices, namely, Contrast, Dissimilarity, Homogeneity, ASM, Energy, and Correlation. Out of these ASM might be dropped, as Energy is directly related to ASM. These five parameters are calculated for each of the four angles (0, 45, 90, 135), thus creating 20 features. The function of *sklearn* that helps us extract the GLCM features is **graycomatrix**. The following code finds four GLCM features of an image called gray_image.

Code:

```
glcm_matrix = graycomatrix(gray_image, distances=[1], angles=[0], levels=256)
contrast_feat=graycoprops(glcm_matrix , 'contrast')
dissimilarity_feat=graycoprops(glcm_matrix , 'dissimilarity')
homogeneity_feat=graycoprops(glcm_matrix , 'homogeneity')
energy=graycoprops_feat(glcm_matrix , 'energy')
correlation_feat=graycoprops(glcm_matrix , 'correlation')
```

Another example of histogram features is Gray-Level Run Length Matrix (GLRL).

Local Binary Pattern

LBP evaluates the weighted average of each pixel followed by the formation of a histogram of the pixel intensities of the so-formed image. It has many variants, the most popular of which are default, ror, nri_uniform, and uniform. The case study given in this chapter describes this feature extraction method in detail. Note that the radius from the central pixel and the number of neighbors are the two most important parameters of this method. Figure 1-10 shows the application of LBP on the image shown in Figure 1-9. The LBP is applied with radii 1 and 2 and the numbers of neighbors 4 and 8 and the methods default, ror, nri_uniform, and uniform.

Code:

```
from matplotlib import pyplot as plt
from skimage.feature import local_binary_pattern
img_arr= plt.imread('spidy.png')
img_arr = img_arr[:,:, 0]
img_lbp_41 = local_binary_pattern(img_arr, 4, 1)
plt.imshow(img_lbp_41)
img_lbp_41_ror = local_binary_pattern(img_arr, 4, 1, method = 'ror')
plt.imshow(img_lbp_41_ror)
img_lbp_41_uniform = local_binary_pattern(img_arr, 4, 1, method='uniform')
plt.imshow(img_lbp_41_uniform)
```

```
img_lbp_41_nri_uniform = local_binary_pattern(img_arr, 4, 1, method='nri_
uniform')
plt.imshow(img_lbp_41_nri_uniform)
```

In the same way, the LBP of the given image with parameters r = 2 and neighborhood = 8 can be found using various versions of LBP.

Figure 1-9. *Original image*

CHAPTER 1 REVISITING MACHINE LEARNING

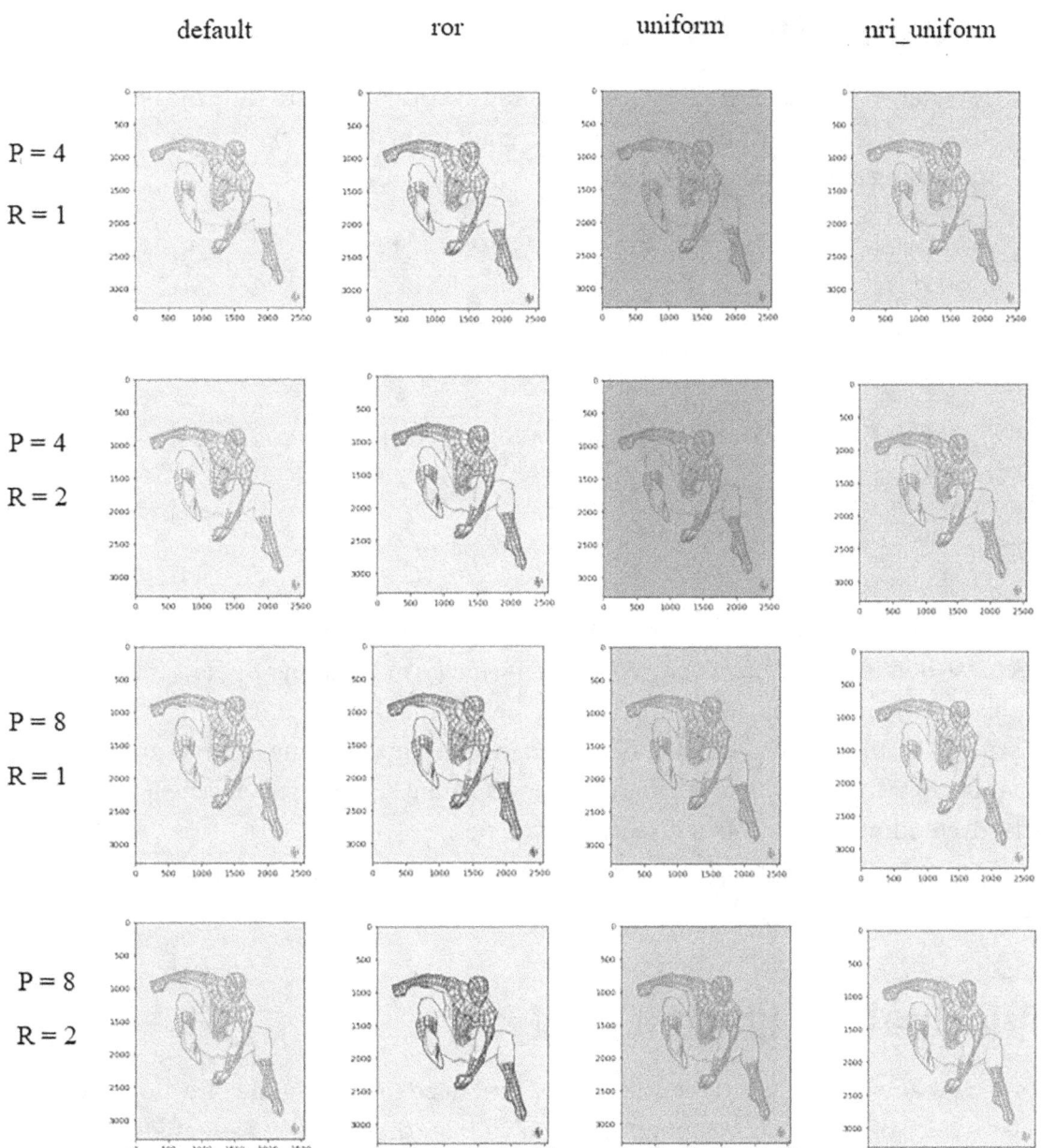

Figure 1-10. *Output: LBP variants with P = 4, 8 and R = 1, 2*

Let us now move to another feature extraction technique called Histogram of Oriented Gradients.

CHAPTER 1　REVISITING MACHINE LEARNING

Histogram of Oriented Gradients

In Histogram of Oriented Gradients, we generally take a block and slide the block over the whole image. For each patch, we find the gradient of that block. These two values can be found using the following formula:

$$H = I(i,j+1) - I(i,j-1)$$

$$V = I(i+1,j) - I(i-1,j)$$

$$Magnitude = \sqrt{(H^2 + V^2)}$$

$$Theta = \left(\frac{V}{H}\right)$$

This is followed by the creation of a histogram of various gradients. The feature vector so obtained can effectively represent the image in terms of oriented gradients.

There are many more feature extraction methods, and only some of them have been discussed in this chapter. It is difficult to find the most appropriate feature extraction method that works for the task at hand. Deep Learning comes to the rescue here, as it effectively eliminates the need of feature extraction.

Let us now have a look at an important feature transformation method called Principal Component Analysis.

Principal Component Analysis

Assume that you have two-dimensional data and need to find out the direction in which the variance of the data is maximum. Assume initially the data is represented in an $X - Y$ coordinate system and this direction turns out to be M. Now the direction that is perpendicular to M, say N, along with M forms the new axis system in which the original data can be transformed and is most probably not correlated.

Principal Component Analysis finds the set of new axes referred to as Principal Directions in which the variation of the data is maximum. This can also be used to reduce the dimensionality of the data. These principal components can be found by finding the eigenvalues and the corresponding vectors from the data covariance matrix. The data covariance matrix can be found by using the following formula:

$$\Sigma = (X - \overline{X})^T \times (X - \overline{X})$$

To find the principal component for X

1. Find the eigenvalues and corresponding eigen data vectors of the covariance matrix Σ.

2. Arrange the eigenvalues in the decreasing order and do the corresponding vectors.

3. The so-arranged eigen vectors are then stacked as *eigen _ vectors*. Note that you can take the requisite number of eigen vectors.

Now find

$$X_{transformed} = X \times eigen\_vectors$$

The shapes of the matrices formed in this process are as follows:

Matrix	Shape
X	$n \times m$
$(X - \overline{X})$	$n \times m$
Σ	$m \times m$
$X_{transformed}$	$n \times m$

The following code implements PCA. Note that the image has been reconstructed using just one principal component, 10 components and 80 components. The output is shown in Figure 1-11.

Code:

```
#Importing Libraries
from matplotlib import pyplot as plt
import numpy as np
from numpy import linalg as LA
```

CHAPTER 1 REVISITING MACHINE LEARNING

```
#Loading image
img1 = plt.imread('Spidy.jpg')
plt.imshow(img1)
def RGBtoGray(img1):
    img_gray = 0.299*img1[:,:,0] + 0.587*img1[:,:,1] + 0.114*img1[:,:,2]
    return img_gray
print(img1.shape)
img_gray = RGBtoGray(img1)
X_mean = np.mean(img_gray, axis=1)
print(X_mean.shape)
X = img_gray
print(X.shape)
X_mean = np.reshape(X_mean, (X_mean.shape[0], 1))
diff = (X- X_mean)
cov1 = np.matmul((X - X_mean).T, (X - X_mean))
print(cov1.shape)
eigenvalues, eigenvectors = LA.eig(cov1)
#print(eigenvalues)
print(eigenvectors.shape)
# 0 Principal Components
T1 = eigenvectors[:,0]
T1 = np.reshape(T1, (T1.shape[0], 1))
print(T1.shape)
Transformed = np.matmul(X, T1)
print(Transformed.shape)
recon = np.matmul(Transformed, T1.T)
print(recon.shape)
plt.imshow(recon)
eigenvalues, eigenvectors = LA.eig(cov1)
#print(eigenvalues)
print(eigenvectors.shape)
# 10 Principal Components
T1 = eigenvectors[:,:10]
#T1 = np.reshape(T1, (T1.shape[0], 1))
print(T1.shape)
```

```
Transformed = np.matmul(X, T1)
print(Transformed.shape)
recon = np.matmul(Transformed, T1.T)
print(recon.shape)
plt.imshow(recon)
eigenvalues, eigenvectors = LA.eig(cov1)
#print(eigenvalues)
print(eigenvectors.shape)
# 80 Principal Components
T1 = eigenvectors[:,:80]
#T1 = np.reshape(T1, (T1.shape[0], 1))
print(T1.shape)
Transformed = np.matmul(X, T1)
print(Transformed.shape)
recon = np.matmul(Transformed, T1.T)
print(recon.shape)
plt.imshow(recon)
```

Output:

The output is shown in Figure 1-11.

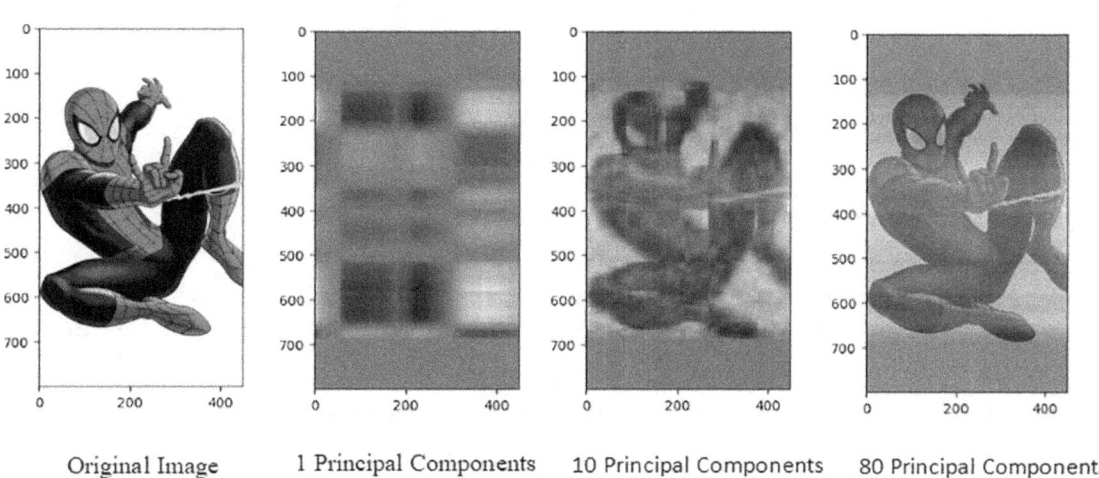

Figure 1-11. Output of the above PCA code

Now, let us move to one of the most important topics in Machine Learning: the bias–variance trade-off.

CHAPTER 1 REVISITING MACHINE LEARNING

Bias–Variance Trade-off

This is perhaps one of the most important topics in Machine Learning. So far we have seen how to reduce the error on the training set using gradient descent. That is, what should be the parameters of the model so as to have minimum training error? However, what matters is the test error, or how well a classifier (or regression algorithm) performs on the test set, that is, how well can it generalize. Let us try to understand the decomposition of this error. Assume that you have a dataset

$$D = \{(x_1, y_1), \ldots, (x_n, y_n)\}$$

drawn from a distribution $\zeta(x, y)$, where $y \epsilon R$ (regression setting). Here, $\zeta(x, y)$ is the probability distribution from which n independent samples have been drawn to create D. Note that $\zeta(x, y) = \zeta(y/x)\zeta(x)$ and $\overline{y}(x)$ is the predicted value of the label y.

We train the ML algorithm M with the training dataset and come up with the hypothesis h on dataset D

$$h_D = M(D)$$

The expected test error in this case will be

$$E = \left[(h_D(x) - y)^2 \right]$$

Based on this error, we find if the model performs well or not.

Overfitting and Underfitting

The Machine Learning model should give a good performance with both the train and the test set. If the model does not perform well with the train set, you can opt for options like having more data hyperparameter tuning or selecting a different learning algorithm.

Overfitting is a condition wherein the model performs well on the train set but poorly on the test set. A complex model generally overfits. In case of overfitting one may opt for the following options.

Bias and Variance

The average prediction of a good Machine Learning model should be as close as the ground truth as possible. This difference is referred to as bias. This can be perceived as the ability of the underlying model to predict values. The formal definition of bias is as follows:

$$Bias = E\left[f'(x) - f(x)\right],$$

where $f(x)$ is the average predicted value of the model and $f(x)$ is the underlying function. High bias indicates the inability of the model to fit the training data. One of the reasons of this may be an oversimplified model. High bias leads to more error rate both with the train and the test set.

The variance of a model signifies its ability to adjust to a given dataset. This variability is referred to as variance. The formal definition of variance is as follows:

$$Variance = E\left[f'(x) - f(x)\right]^2,$$

High variance may be due to the model being too complex. An overcomplex model may lead to low error with the train set but a high error with the test set.

Ideally, one should plot a graph of the variation of bias and variance with the iterations. Note that the bias should decrease with the iterations, whereas the variance may increase after a point. The aim is to look for a point where both these curves meet.

Figure 1-12 shows the four possibilities vis-à-vis the bias and variance. Figure 1-12 (a) shows the case with low bias and low variance (ideal). In this case both the training and the test performance are the same. Figure 1-12 (b) shows the case of low bias and high variance wherein the performance of the model with the train set is fair, but with the test set may be poor. In the case of high bias and low variance (Figure 1-12(c)), the train performance may not be good, but the difference between the performance of the model for the train and the test set may not be huge.

It may be noted that bias and variance and underfitting and overfitting are closely related, as discussed.

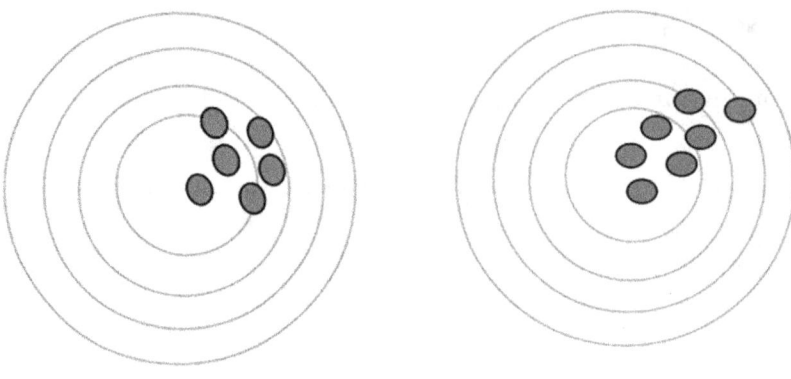

1-12 (a). Low Bias, Low Variance *1-12 (b). Low Bias, High Variance*

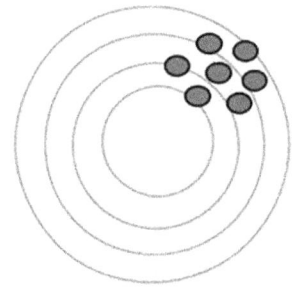 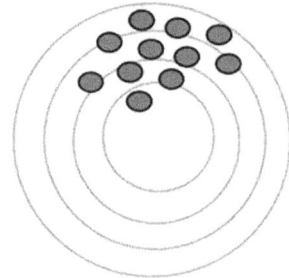

1-12 (c). High Bias, Low Variance *1-12 (d). High Bias, High Variance*

Figure 1-12. *Bias-variance*

The concept has been discussed in detail in the Chapter 5 on hyperparameter tuning. As a matter of fact, handling the bias and the variance forms an essential part of the development of any successful ML or Deep Learning model.

Application: Classification of Handwritten Digits Using a Conventional Machine Learning Pipeline

As discussed in the previous sections, the Machine Learning pipeline consists of preprocessing, feature extraction, feature selection, learning, and post-processing. This section explores the classification of the MNIST dataset and applies various feature selection and extraction methods and compares the results using three different types of classifiers:

CHAPTER 1 REVISITING MACHINE LEARNING

K-Nearest Neighbors (KNN), Neural Networks, and Support Vector Machine (SVM). KNN and SVM have already been discussed; Chapter 3 discusses NN in detail.

Dataset

The MNIST dataset is a widely used dataset, consisting of 70,000 images of handwritten digits from 0 to 9, each of size 28 × 28 pixels. The training set consists of 60,000 images, and the test set contains 10,000 images.

Data Preprocessing

The dataset consists of grayscale images of size 28 × 28 pixels, having pixel values between 0 and 255. The LBP replaces each pixel of the given image by the weighted average of its neighbors. For example, in the following 10 × 10 image, the central pixel is taken as reference, and its eight neighbors are considered. The cells having pixel value greater than the reference are then replaced by 1, and those having less than the reference are replaced by 0. The binary number so formed by traversing the neighbors is then converted into a decimal number, and then the reference pixel is replaced by this value (Figure 1-13).

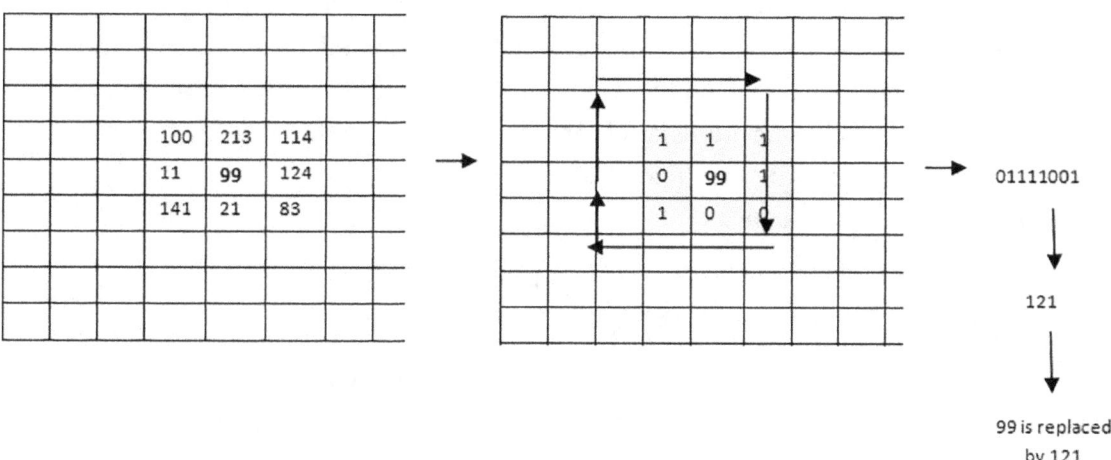

Figure 1-13. Computing LBP

The process is repeated for all the pixels in the given image. The application of LBP on an image results in the formation of a new image having edges. Figure 1-14 (b) shows the resultant image, when LBP is applied on the image shown in Figure 1-14 (a).

CHAPTER 1   REVISITING MACHINE LEARNING

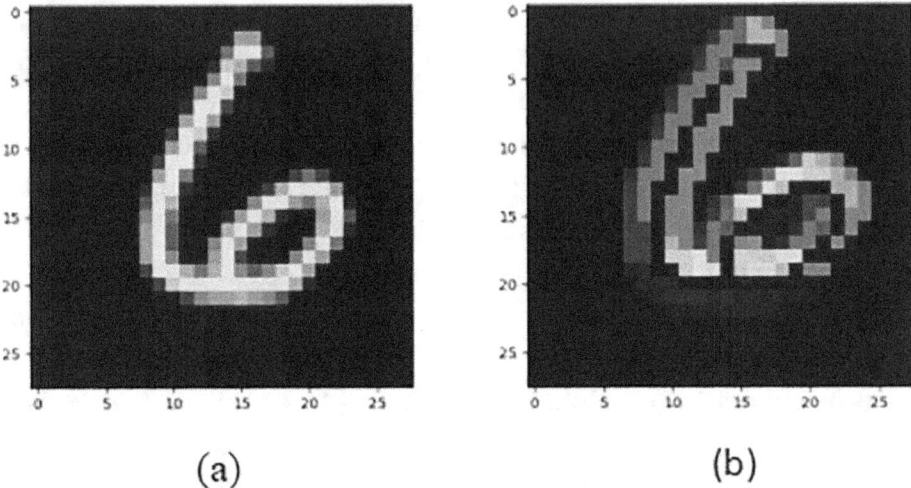

Figure 1-14. *Application of LBP on an image*

The frequency of each pixel in the image so formed is then determined. This feature extraction method (referred to as FE1 in this section) has three variants: default, rotation-invariant, and uniform rotation-variant. LBP is applied to each image, and the resulting features are concatenated vertically to create the feature matrix "X." Simultaneously, the corresponding labels are stored in a variable "y." The dataset is then split into training and testing sets in a 90:10 ratio.

Feature Extraction Variants

>Default LBP Variant: This variant captures the local texture patterns of each digit in its original form.

>Rotation-Invariant LBP Variant: This variant ensures that the extracted features remain consistent even if the digits undergo rotational transformations.

>Uniform Rotation-Variant LBP Variant: This variant focuses on uniform patterns, providing a more robust representation of digit textures.

Feature Selection

To enhance model performance and reduce dimensionality, feature selection is performed using the F1 method. This step aims to retain the most informative features while eliminating redundant or irrelevant ones.

Classification Algorithms

Three classification algorithms (On the three variants of LBP) denoted as C1, C2, and C3 are employed to predict the digit labels based on the extracted and selected features.

Performance Evaluation

Performance evaluation is conducted using the F-measure metric, considering both macro- and micro-average values.

Code:

```
#Importing Libraries
import tensorflow as tf
import keras
from matplotlib import pyplot as plt
import numpy as np
from numpy import genfromtxt
from sklearn.svm import SVC
from sklearn.tree import DecisionTreeClassifier
from sklearn.neural_network import MLPClassifier
from sklearn.neighbors import KNeighborsClassifier
from sklearn.metrics import f1_score
from sklearn.metrics import confusion_matrix
from sklearn.multiclass import OneVsRestClassifier
from sklearn.feature_selection import VarianceThreshold

#Loading Dataset
tf.keras.datasets.mnist.load_data(path="mnist.npz")
#Train Test Split
(X_train, y_train), (X_test, y_test) = keras.datasets.mnist.load_data()
print(X_train.shape, y_train.shape, X_test.shape, y_test.shape )
#Local Binary Pattern
def LocalBinaryPattern(img1):
    result1 = np.zeros((img1.shape[0], img1.shape[1]))
    for i in range(1, img1.shape[0]-1):
        for j in range(1, img1.shape[1]-1):
            val = [0]*8
            val[0] = img1[i, j-1]>img1[i,j]
            val[1] = img1[i-1, j-1]>img1[i,j]
            val[2] = img1[i-1, j]>img1[i,j]
```

```python
                    val[3] = img1[i-1, j+1]>img1[i,j]
                    val[4] = img1[i, j+1]>img1[i,j]
                    val[5] = img1[i+1, j+1]>img1[i,j]
                    val[6] = img1[i+1, j]>img1[i,j]
                    val[7] = img1[i+1, j-1]>img1[i,j]
                    sum1 = 0
                    for k in range(8):
                            sum1+= val[k]*(2**k)
                    result1[i, j]= sum1
        return result1
def LBP_Feat(LBP_image):
        feat= [0]*256
        num, count1 = np.unique(LBP_image, return_counts=True)
        LBP_Features1 = dict(zip(num, count1))
        for i in range(256):
                if i in LBP_Features1:
                        feat[i]= LBP_Features1[i]
                else:
                        feat[i]= 0
        return feat

#Applying Local Binary Pattern on X_train and X_test
def CreateX(X_images):
        X = np.zeros((1, 256))
        for i in range(X_images.shape[0]):
                image1 = X_images[i, :, :]
                LBP_image = LocalBinaryPattern(image1)
                feat = LBP_Feat(LBP_image)
                feat1 = np.reshape(feat, (1, 256))
                X = np.vstack((X, feat1))
                if(i%100 == 0):
                        print('Iteration ',i)
                X= X[1:,:]
        return (X)

X_train = CreateX(X_train)
np.savetxt("X_train_MNIST_LBP.csv", X_train, delimiter=",")
```

```python
X_test = CreateX(X_test)
np.savetxt("X_test_MNIST_LBP.csv", X_train, delimiter=",")

#Training Model with KNN
#KNN with K = 5
clf = KNeighborsClassifier(n_neighbors=5)
clf.fit(X_train, y_train)
y_predict = clf.predict(X_test)
confusion_matrix(y_test, y_predict)

#KNN with K=3
clf = KNeighborsClassifier(n_neighbors=3)
clf.fit(X_train, y_train)
y_predict = clf.predict(X_test)
confusion_matrix(y_test, y_predict)
#Plotting F score of KNN-3 and KNN-5 for all the classes
import matplotlib.pyplot as plt
plt.style.use('seaborn-deep')
X_axis = np.arange(len(KNN3_F_Score))
plt.bar(X_axis - 0.2, KNN3_F_Score, 0.4, label = 'KNN3')
plt.bar(X_axis + 0.2, KNN5_F_Score, 0.4, label = 'KNN5')
X_labels = ['Class'+str(i) for i in range(1, 11)]
plt.xticks(X_axis, X_labels)
plt.xlabel("Model")
plt.ylabel("F Score")
plt.title("Comparison of KNN3 and KNN5")
plt.legend()
plt.show()

#Training Model with Decision Tree
clf = DecisionTreeClassifier(random_state=0)
clf.fit(X_train, y_train)
y_predict = clf.predict(X_test)
confusion_matrix(y_test, y_predict)
DT_F_Score= f1_score(y_test, y_predict, average=None)

#Plotting Performance of KNN-5 and DT
```

CHAPTER 1   REVISITING MACHINE LEARNING

```
X_axis = np.arange(len(DT_F_Score))
plt.bar(X_axis - 0.2, KNN5_F_Score, 0.4, label = 'KNN3')
plt.bar(X_axis + 0.2, DT_F_Score, 0.4, label = 'DT')
X_labels = ['Class'+str(i) for i in range(1, 11)]
plt.xticks(X_axis, X_labels)
plt.xlabel("Model")
plt.ylabel("F Score")
plt.title("Comparison of KNN5 and DT")
plt.legend()
plt.show()
```

Results: The model is implemented and the results are observed. The reader is expected to run the above code and observe the performance measures in the following cases:

Without Feature Selection (P1, P2, P3): The classification algorithms should initially be applied to the raw feature matrix "X" without feature selection, and the following results should be noted:

- P1 (C1 without feature selection)
- P2 (C2 without feature selection)
- P3 (C3 without feature selection)

After Feature Selection (P11, P22, P33): The same classification algorithms should then be applied after feature selection using the F1 method, and the following results should be noted:

- P11 (C1 after feature selection)
- P22 (C2 after feature selection)
- P33 (C3 after feature selection)

Compare your results with the following outputs.
Output:

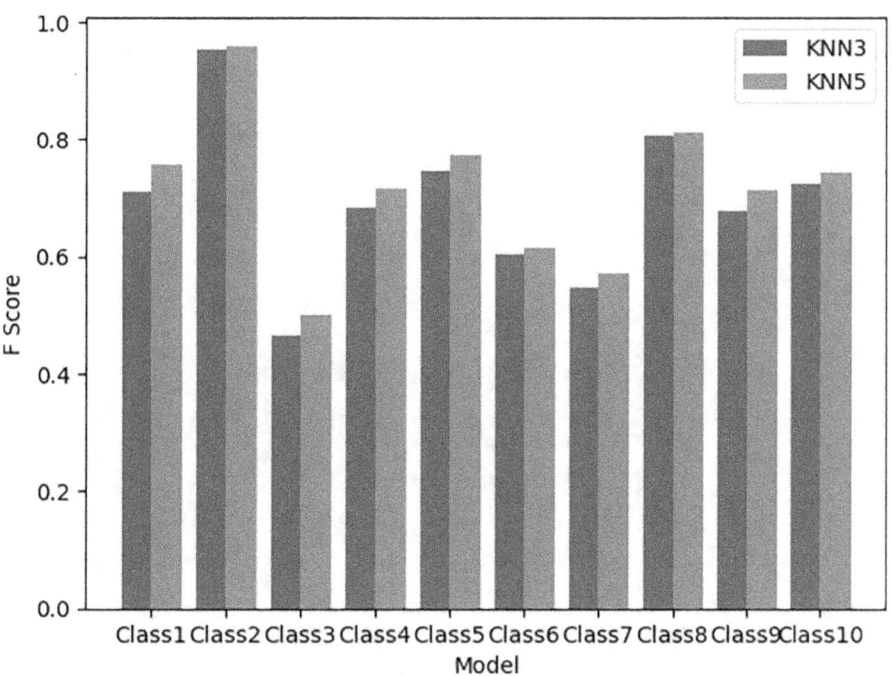

CHAPTER 1   REVISITING MACHINE LEARNING

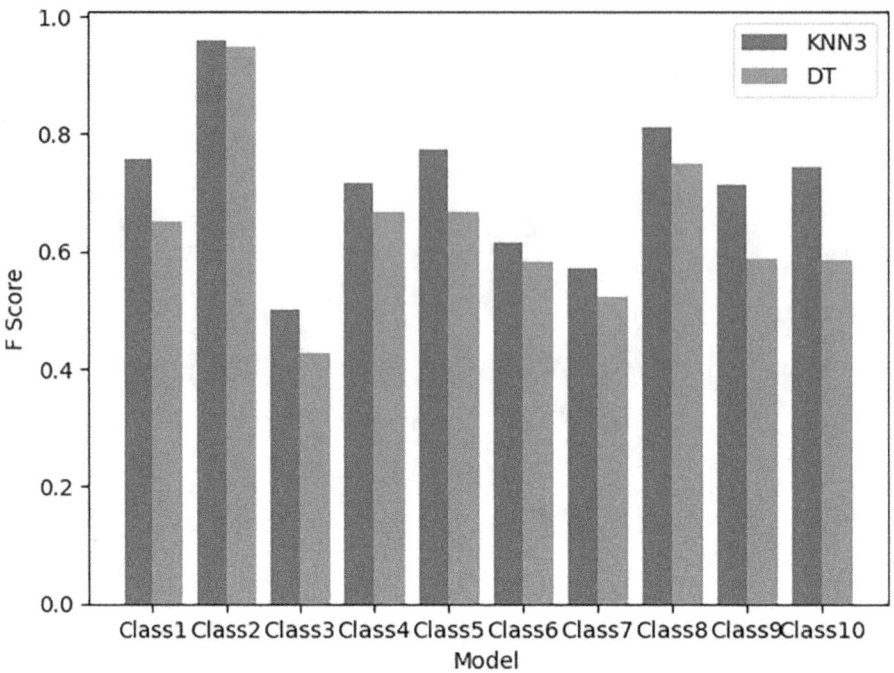

The reader is expected to analyze the results and figure out why a particular combination works well for this dataset. If you find it difficult, you may refer to Appendix A.

Conclusion

This chapter introduces Machine Learning and discusses its evolution and types. The chapter also hovers the feature extraction and feature selection methods. It then discusses a detailed pipeline that allows for a thorough exploration of the impact of different feature extraction variants, feature selection, and classification algorithms on the task of handwritten digit classification (case study in the previous section). The results obtained from the various combinations of these techniques will provide insights into the effectiveness of the proposed pipeline and aid in selecting the most suitable approach for this specific problem. Now that you know that it is difficult to select the best feature extraction and selection method for your problem and that a lot of effort is required to handle the bias and variance, let us move to Deep Learning. The next chapter introduces Deep Learning.

Exercises
Multiple-Choice Questions

1. Which of the following can be used to extract features from an image?

    a. Local Binary Pattern

    b. Histogram of Oriented Gradients

    c. Gray-Level Co-occurrence Matrix

    d. All of the above

2. Which of the following finds the weighted average of the neighborhood pixels in an image and then creates a histogram of pixel intensities?

    a. Local Binary Pattern

    b. Histogram of Oriented Gradients

    c. Gray-Level Co-occurrence Matrix

    d. All of the above

3. In which of the following a matrix depicting the occurrence of a gray-level value near another is formed?

    a. Local Binary Pattern

    b. Histogram of Oriented Gradients

    c. Gray-Level Co-occurrence Matrix

    d. All of the above

4. We should not use the raw pixels as features in a binary classification problem having a dataset of 60 images (of size 1024 × 1024), consisting of two classes. Why?

    a. Curse of dimensionality

    b. Memory requirement

    c. Computation time

    d. All of the above

CHAPTER 1 REVISITING MACHINE LEARNING

5. You have a labeled dataset having 10 features and 100 rows. You need to reduce the dimensionality or transform features to improve performance. Which of the following cannot be used for this purpose?

    a. Local Binary Pattern

    b. PCA

    c. FDR

    d. Wrapper methods

6. Can you represent a 1024 × 1024 image in terms of a feature vector having 128 bins?

    a. Yes

    b. No

7. Which of the following is a filter method?

    a. FDR

    b. RFE

8. Which of the following is a wrapper method?

    a. FDR

    b. RFE

9. If a model does not perform well even on a training set, it suffers from …?

    a. High bias

    b. Low bias

    c. High variance

    d. Low variance

10. If a model performs well even on a training set, but poorly on the test set, it suffers from …?

    a. High bias

    b. Low bias

    c. High variance

    d. Low variance

Applications

Collect 50 pictures of Bart Simpson from the popular cartoon *The Simpsons*. Also collect 50 pictures of Homer from the same series. Perform the following tasks on the so-collected images:

1. Reshape all the images into 100 × 100 images, using Python.

2. Now extract features from both classes using all three variants of Local Binary Pattern.

3. Use the above features to classify the two classes, with and without the following feature extraction methods:

    a. Fisher Discriminant Ratio

    b. Recursive Feature Elimination using SVM

    c. Recursive Feature Elimination using Decision Tree

4. Report the performance in each case and discuss why some combinations work better than others.

5. Perform the above tasks (Q3 and Q4) using features obtained from Gray-Level Co-occurrence Matrix.

6. Perform the above tasks (Q3 and Q4) using features obtained from Histogram of Oriented Gradients.

7. Carry out experiments to find if the application of Principal Component Analysis on the features obtained in Q2, Q5, and Q6 improves the performance.

8. Perform linear regression on the Boston Housing price dataset, after selecting features using RFE.

References

[1] Bishop, C. M. (2006). Pattern recognition and machine learning. Springer Verlag.

[2] Goodfellow, I., Bengio, Y., & Courville, A. (2016). Deep learning. MIT Press.

[3] Mitchell, T. M. (1997). Machine learning.

[4] Canny J (2024) Introduction To Data Science Unsupervised Learning.

[5] Bhasin, H. (2023). Machine Learning for Beginners: Build and deploy Machine Learning systems using Python - 2nd Edition. BPB Publications.

CHAPTER 2

Introduction to Deep Learning

Neurons

The study of how the brain works has fascinated scientists for long. This fascination got the wings with the advent of histology, unveiling how neurons are organized. The neuron doctrine states that the nervous system comprises independent neurons. However, earlier it was widely believed that the nervous system consists of a single continuous network, a theory proposed by Joseph von Gerlach and propounded by Camillo Golgi. Golgi also invented a staining technique that helped prove this theory wrong. Using this technique, Santiago Ramón y Cajal proved that the neurons are independent cells. The structure of neurons was also unveiled using this very staining technique. The term *neuron* was proposed by Heinrich Wilhelm Gottfried Waldeyer-Hartz in around 1891.

Both Golgi and Cajal (Figure 2-1(a)) were awarded the 1911 Nobel Prize for their work. Electron microscopy later proved that neurons are, in fact, independent cells. These neurons inspired Neural Networks. The cartoon shown in Figure 2-1(b) is made using an application, Imagen, that uses these Neural Networks. The above timeline is shown in Figure 2-1(c).

CHAPTER 2 INTRODUCTION TO DEEP LEARNING

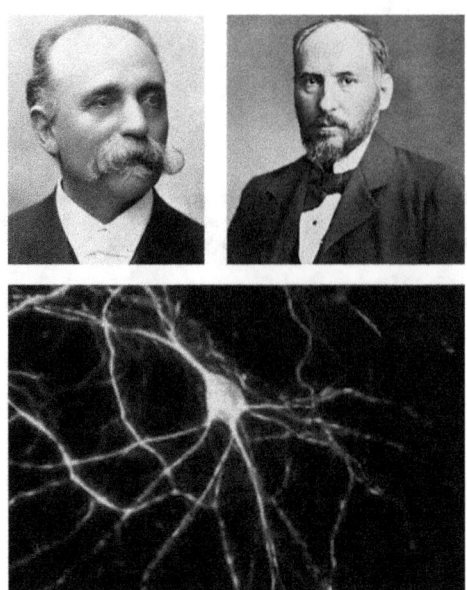

Figure 2-1(a). *Golgi (top left), Cajal (top right), and staining of neurons (bottom)*

Figure 2-1(b). *Golgi, Cajal, and neuron: cartoon made using Imagen, a Deep Learning–based application*

CHAPTER 2　INTRODUCTION TO DEEP LEARNING

TIMELINE

Neuron

1871–1873 RECTICULAR THEORY

STAINING TECHNIQUE

1888–1891 NEWTON DOCTRINE

1950s ELECTRON MICROSCOPE

Figure 2-1(c).   *Timeline: neuron*

From Perceptron to the Winter of Artificial Intelligence

The models that we will discuss in this book are Deep Neural Networks (DNNs). These models are based on Neural Networks, which draw their inspiration from a neuron. The first computational model inspired by a neuron was the McCulloch–Pitts model. This model was proposed by a neurologist, McCulloch, and a logician, Pitts. The model had binary inputs $x_1, x_2, x_3, \ldots, x_n$, $x_i \epsilon \{0, 1\}$, binary output, $y \epsilon \{0, 1\}$, and a thresholding unit. The models were able to implement logic gates, and hence it was established that they could implement a logic machine.

Frank Rosenblatt proposed continuous weights and inputs, which markedly improved the power of perceptrons. Now, they could be used for linear classification and regression. People said that they would be able to rule the world, but the enthusiasm did not last long. Minsky and Papert wrote a book called *Perceptrons*, in which they proved that the perceptrons had limitations. They, in particular, discussed the XOR problem, which could not be solved by these models. This led to the withering of interest in Neural Networks from 1969 to the mid-1980s. In 1986 Rumelhart et al. proposed the backpropagation algorithm that could be used to train a Multi-layer Perceptron (MLP). This helped develop Neural Networks having multiple layers that could solve the XOR problem and greatly improved their performance on various supervised learning tasks. The concept of pretraining the networks was proposed in 1991, which laid the foundation of many works after 2006.

The above timeline is shown in Figure 2-2.

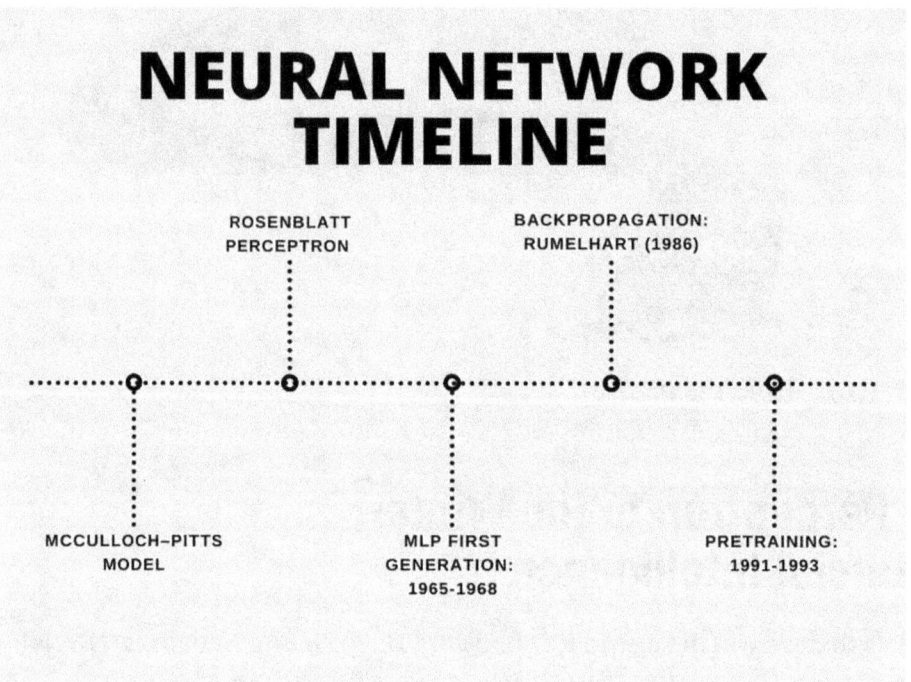

Figure 2-2. *Timeline: Neural Networks*

Chapter 3 of this book discusses all these models in detail.

CHAPTER 2 INTRODUCTION TO DEEP LEARNING

Imagery and Convolutional Neural Networks

For imaging-related tasks, a network capable of inferring the spatial correlation was needed. Cats helped the Scientific Fraternity come up with such networks. (Really!) An experiment on cats proved that only some parts of the brain are activated in response to a particular stimulus (Figure 2-3). The experiment was carried out by Hubel and Wiesel who showed that

> *A Neuron fires only in response to a particular stimulus in a particular region.*

The development of Neural Networks, based on the above concept, called the Convolutional Neural Network (CNN), dates back to the 1990s. A CNN, called LeNet, was proposed to identify handwritten digits. This work also gave us the famous MNIST dataset, which was later used by many scientists, working in this field, to test their model.

Figure 2-3. *Only some parts of a cat's brain are activated on seeing a particular image*

CHAPTER 2   INTRODUCTION TO DEEP LEARNING

The advances in image analysis got a boost with the advent of the ImageNet competition. It had 1000 classes and numerous images. As per the official site

> ***ImageNet*** *is an image database organized according to the WordNet hierarchy, in which each node of the hierarchy is depicted by hundreds and thousands of images. The project has been instrumental in advancing computer vision and deep learning research. The data is available for free to researchers for non-commercial use. [3]*

The models that won or performed well in this competition later became important in this field. Some of these included

- AlexNet, which has eight layers
- ZFNet, which has eight layers but a better error rate as compared with AlexNet
- VGG-Net, which has 19 layers and still a better error rate
- GoogLeNet
- ResNet

We will discuss some of these models in Chapter 7 of this book.

The last two decades witnessed many tasks being accomplished by deep networks. One of the first was handwriting recognition. Graves et al. outperformed the then state-of-the-art models in 2009 and that, too, for Arabic handwriting. Cireşan et al. created a benchmark on the MNIST dataset. The next year saw the advent of a pattern recognizer for the IJCNN traffic sign recognition system.

2016 saw major advances in the Speech Recognition System with an improvement of around 16% vis-à-vis the then state-of-the-art models on a dataset.

The above discussion is summarized in Figure 2-4.

CHAPTER 2   INTRODUCTION TO DEEP LEARNING

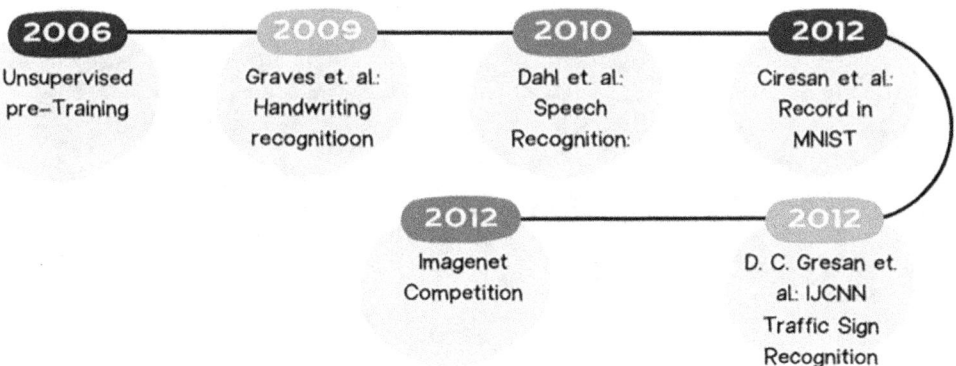

Figure 2-4. *The advent of Deep Learning*

The CNN models are introduced in Chapter 6 of this book.

What's New

The advent of better optimizers led to better convergence and better accuracies. The optimization algorithms starting from gradient descent, Nesterov (1983), AdaGrad (2011), RMSprop (2012), and Adam (2015) set the things rolling for deep networks. As a matter of fact, many new algorithms including Eve and Beyond Adam were proposed later (Figure 2-5).

CHAPTER 2 INTRODUCTION TO DEEP LEARNING

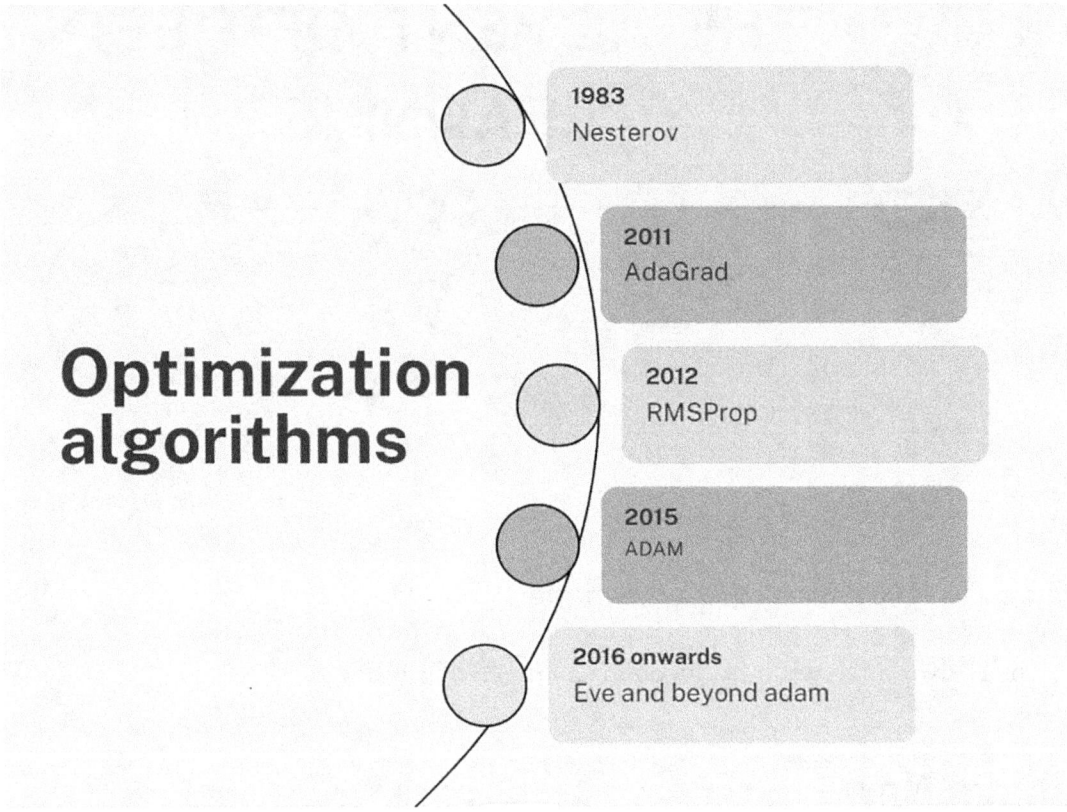

Figure 2-5. Optimization algorithms

The above was accompanied by new activation functions like tanh (1991), Rectified Linear Unit (ReLU) (2010), Leaky ReLU (2013), and SIREN (2020). Also, the betterment in the hardware has contributed to the development of Deep Learning.

Sequences

Though fully connected networks helped us crack many tough problems and Convolutional Neural Networks helped us solve many image-related problems, the problems related to sequences were yet to be handled.

The handling of sequences could solve problems related to text, speech, time series, and so on. In these problems, the relation between the different steps of a sequence plays an equally important role. The Hopfield Network proposed in 1982 modeled a content-addressable memory. The Jordan network gave the idea of having the output of one state become an input to the next state. Likewise, the idea of the hidden state of a network becoming the hidden state of another network was proposed by the Elman network.

The recursive networks so developed suffered from problems like vanishing gradient. The problem was addressed by models like Gate Recurrent Unit (GRU) and Long Short-Term Memory (LSTM).

Chapter 8 of this book introduces these models to the reader.

The above timeline is shown in Figure 2-6.

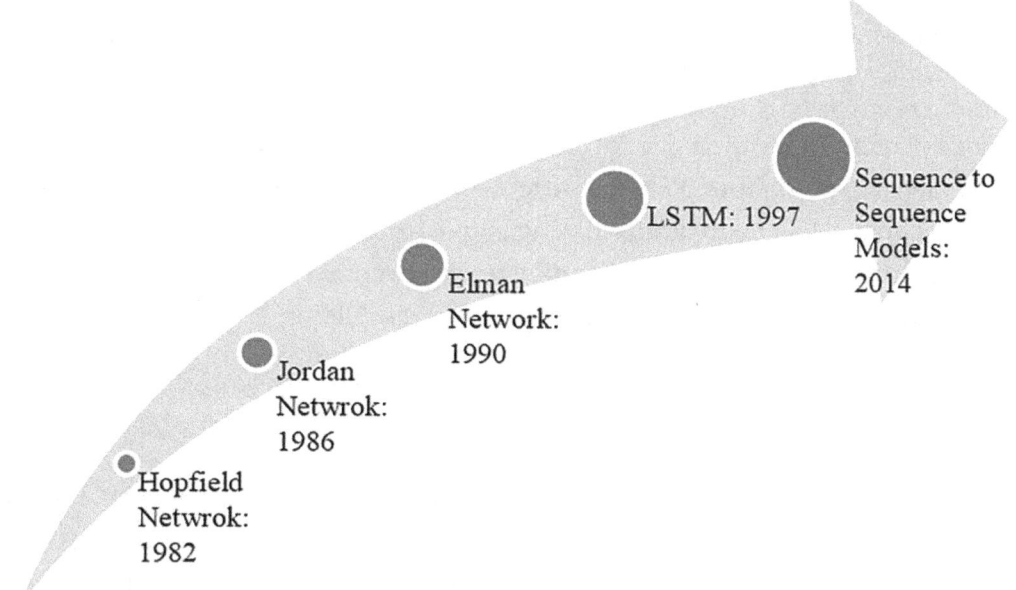

Figure 2-6. Timeline: sequence models

The Definition

The last chapter discussed Machine Learning and its pipeline. To apply Machine Learning for supervised and unsupervised tasks, preprocessing, feature extraction, and feature selection are required. Feature extraction is generally modality specific. Moreover, one can apply numerous feature extraction methods to represent the given data. The selection of the optimal methods is a precarious task. The same is the case with feature selection. As discussed in the previous chapter, feature selection can be done using filter and wrapper methods. So there are numerous techniques for selecting the most pertinent features.

The Deep Learning methods extract the appropriate features and select the most important ones without explicitly stating which one to use.

Moreover, Deep Learning generally results in better performance provided that a sufficient amount of data is given as input to the model. They use state-of-the-art optimization methods and make appropriate use of the hardware. Formally, Deep Learning may be defined as follows:

> *Deep-learning methods are representation-learning methods with multiple levels of representation, obtained by composing simple but non-linear modules that each transform the representation at one level (starting with the raw input) into a representation at a higher, slightly more abstract level. [1]*

Since this training requires a lot of data and resources and most of the time, we do not have such a large amount of data or resources, some models are trained on large datasets by companies and institutes having ample resources, and then they are used to accomplish similar tasks with similar datasets. Here comes the concept of transfer learning. Formally, transfer learning may be defined as follows:

> *Transfer learning is the ability of a system to recognize and apply knowledge and skills gained in previous tasks to new tasks. [2]*

Deep Learning not just extracts the features and selects the pertinent features, but can implement each and every step of the conventional Machine Learning pipeline. This is generally referred to as end-to-end learning. End-to-end learning may be defined as follows:

> *End-to-end learning allows neural networks to transform raw data inputs (such as images) through a series of operations, culminating in final predictions (like class probabilities). This entire transformation process is optimized simultaneously using backpropagation, where the parameters of all layers are adjusted together based on the loss calculated at the output layer. [3]*

Generate Data Using Deep Learning

From classifying digits to writing stories and creating images, we have come a long way. Table 2-1 shows some of the important applications and platforms that help generate text, audio, video, and images using Deep Learning. The reader is expected to explore each of them and access the output. You will get a very good idea of the heights to which the Deep Learning community has scaled over the last two decades.

Table 2-1. *Tools That Use Deep Learning to Generate Text, Audio, Video, and Images*

Modality	Logo	Name	Functionality	URL
Text-to-text Image-to-text		ChatGPT	It allows the user to engage in human-like conversations and accomplish assorted tasks. It can even answer questions and help you in writing text.	https://chatgpt.com/
Text-to-text Image-to-text		Gemini	It can be used to write something new or to rewrite a given piece of text.	https://gemini.google.com/app
Text-to-text Image-to-text Text-to-image Image-to-image Voice-to-text Voice-to-Image		Microsoft Copilot	It also helps in writing, editing, summarizing, and generating content.	https://copilot.microsoft.com/?form=MA13LV#
Text	—	Bert	It is a language model for natural language processing. It can help machines understand the meaning of text using context.	https://huggingface.co/welcome
Text-to-image		Picsart	It converts text into images.	https://picsart.com/ai-image-generator/

(*continued*)

CHAPTER 2　INTRODUCTION TO DEEP LEARNING

Table 2-1. (*continued*)

Modality	Logo	Name	Functionality	URL
Text-to-image		Canva	It lets you choose image variations based on a preferred look and composition.	https://www.canva.com/ai-image-generator/
Text-to-image		Adobe	It lets us generate images from text.	https://www.adobe.com/products/firefly/features/text-to-image.html
Text-to-speech	IIElevenLabs	ElevenLabs	Create natural AI voices in any language.	https://elevenlabs.io/
Text-to-speech		PlayHT	Ultra-realistic text-to-speech (TTS) voice. Leading AI voice generator.	https://play.ht/
Text-to-video	invideo	Invideo AI	Invideo AI serves as an AI video generator that takes your input script and intelligently crafts it into a video.	https://invideo.io/make/add-text-to-video-online/

Conclusion

This chapter presented a brief overview of the timeline of Deep Learning. The development of the first computational model inspired by the structure of a neuron, the McCulloch–Pitts model, and the present-day generative models have been introduced in this chapter. The chapter, in particular, deals with neurons, Neural Networks, Convolutional Neural Networks, sequence models, and the latest tools currently being used to accomplish an assortment of tasks, from writing letters to generating images. The models will be discussed, in detail, in the following chapters. Machines are becoming creative, and they will become more creative in the days to come. We conclude our discussion with the story generated by a Large Language Model (LLM) called Gemini. Have a look at the output (Figure 2-7)! If you find it interesting, then this book is a first step toward making you capable of writing programs that can generate interesting things. Welcome to the world of Deep Learning!

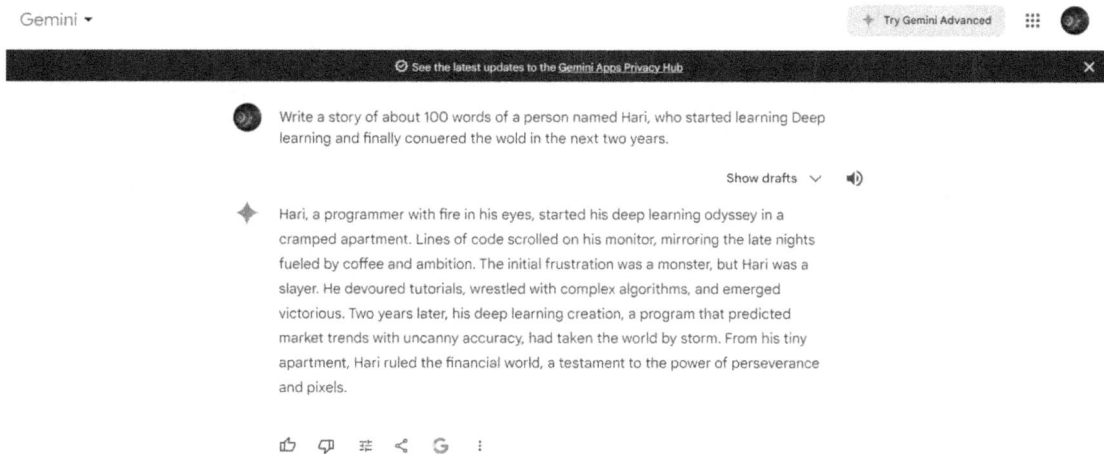

Figure 2-7. *Story generated by Gemini*

CHAPTER 2 INTRODUCTION TO DEEP LEARNING

Exercises
Multiple-Choice Questions

1. Who proposed the structure of a neuron?

    a. Cajal

    b. Golgi

    c. Heinrich Wilhelm Gottfried

    d. None of the above

2. Who proposed the staining technique that helped unveil the structure of a neuron?

    a. Cajal

    b. Golgi

    c. Heinrich Wilhelm Gottfried

    d. None of the above

3. The 1911 Nobel Prize was awarded to

    a. Cajal

    b. Golgi

    c. Both Cajal and Golgi

    d. Heinrich Wilhelm Gottfried

4. The nervous system contains independent neurons. This is

    a. The neuron doctrine

    b. Reticular theory

    c. None of the above

5. The nervous system contains a single continuous network. This is

    a. The neuron doctrine

    b. Reticular theory

    c. None of the above

6. Which was one of the first computational models inspired by a neuron?

    a. McCulloch–Pitts model

    b. Rosenblatt Perceptron

    c. Multi-layer Perceptron

    d. None of the above

7. Which of the following had binary input and binary outputs?

    a. McCulloch–Pitts model

    b. Rosenblatt Perceptron

    c. Multi-layer Perceptron

    d. None of the above

8. Which of the following had continuous input and corresponding weights that could change?

    a. McCulloch–Pitts model

    b. Rosenblatt Perceptron

    c. Multi-layer Perceptron

    d. None of the above

Activity

1. Explore how the working of the nervous system inspired the computing fraternity. Write a short note of about 100 words.

2. To accomplish the above task, you can take the help of references given at the end of the book. Now, draw some infographics to make your article interesting.

3. Now use any publicly available pretrained Large Language Model to write the above article. Compare your article with that generated by Deep Learning.

4. Generate images for the same using GenAI tools available on the Internet.

5. Now search for a research paper published this year on Depression Detection Using Genome data. Read the abstract and make notes.

    Using these notes, ask a pretrained LLM to generate an article.

6. Do you think that the model is able to generate an article of the same quality as in the earlier case? If not, why?

References

[1] LeCun, Y., Bengio, Y. & Hinton, G. Deep learning. *Nature* 521, 436–444 (2015). https://doi.org/10.1038/nature14539

[2] Coulibaly, S., Kamsu-Foguem, B., Kamissoko, D., & Traore, D. Deep Convolution Neural Network sharing for the multi-label images classification. *Machine Learning With Applications* 10, 100422 (2022). https://doi.org/10.1016/j.mlwa.2022.100422

[3] *Stanford University CS231N: Deep Learning for Computer Vision* (n.d.). https://cs231n.stanford.edu/

CHAPTER 3

Neural Networks

Objectives

After reading the chapter, the reader will be able to

- Understand Single-Layer Perceptron.
- Understand the XOR problem.
- Learn about activation functions.
- Appreciate the concept, algorithm, and implementations of Multi-layer Perceptron.
- Understand how Multi-layer Perceptron can solve the XOR problem.
- Learn the backpropagation algorithm.

Introduction

Our brain receives signals via neurons, processes them, and generates responses. Generally, the receptors send information to the neurons, which is passed to the brain. The brain, in turn, processes these signals and sends the response to the effectors. This concept was given by Cajal [1]. Though these neurons are slower than the logic gates, their magnitude helps us deal with the given situation quickly.

The structure of a neuron is shown in Figure 3-1. The dendrites act as receptor zones, the cell body processes the inputs, and the axons transmit the signals. The neurons are connected to each other via synapses.

CHAPTER 3   NEURAL NETWORKS

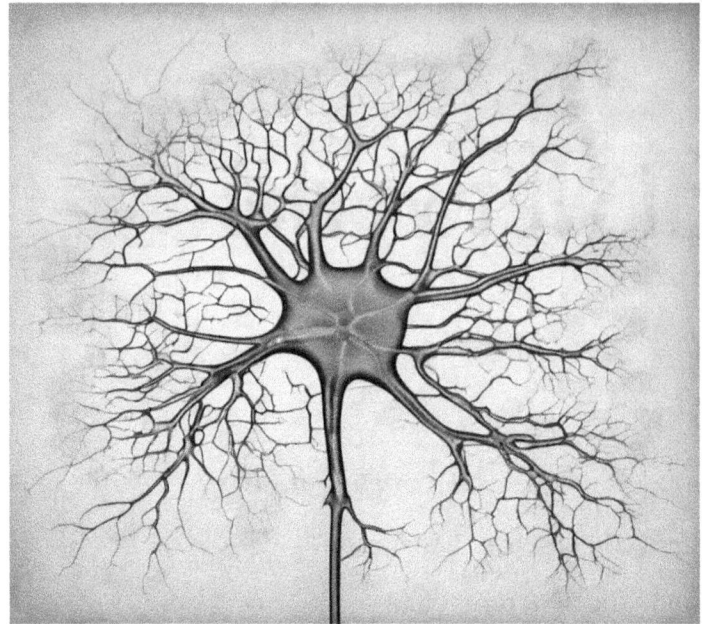

Figure 3-1. *Picture of a neuron generated using AI (`https://pixlr.com/image-generator/`)*

The computational model shown in Figure 3-2 is similar to the neuron. This model receives a two-dimensional input and classifies the input into one of the two classes.

It receives the inputs (X_1, X_2) from the input nodes, multiplies them with the corresponding weights (W_1, W_2), takes the summation, and passes it through a function. If the output of this function is greater than the threshold, then the output of the model becomes 1; else, it becomes 0. This model can therefore act as a binary classifier.

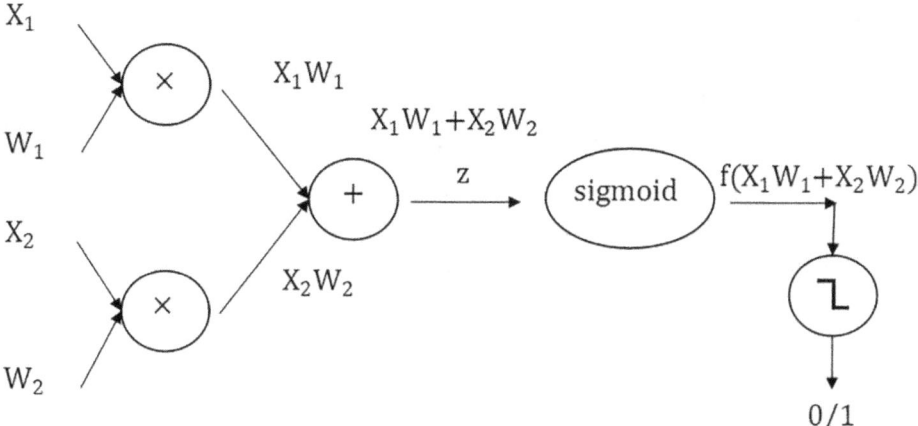

Figure 3-2. *Computational model based on the structure of a neuron*

The above model, referred to as Single-Layer Perceptron or SLP, can be extended to one that takes "d" inputs. The following points are worth noting as regards SLP:

- The number of neurons in the input layer is the same as the number of inputs.

- The number of weights will be the same as the number of inputs, and each weight denotes the importance of that input.

- The weighted sum presents the linear combination of the inputs and the weights.

- The weighted sum added with the bias passes through an activation function. The activation functions are discussed in section "Activation Functions."

If the output of the last step is greater than the threshold, then the final output of the model is 1; else, it is 0.

The above model is referred to as the Rosenblatt Perceptron model named after Frank Rosenblatt [2]. Now consider a simpler model in which the inputs are binary (either 0 or 1) and each of the weights has values that signify the importance of the input. The weights can be positive or negative, depicting excitatory or inhibitory connections. This model is referred to as the McCulloch and Pitts model [3]. This model can be used to implement logic gates, wherein output can be classified using a linear hyperplane. Having seen the basics, let us now move to a detailed discussion on SLP.

CHAPTER 3 NEURAL NETWORKS

Single-Layer Perceptron

A Single-Layer Perceptron is a linear classifier. It can classify two classes that can be segregated using a line in the case of two dimensions, a plane in the case of three dimensions, and a hyperplane in the case of multiple dimensions. However, it cannot classify nonlinearly separable data. Let's discuss how this classification can be done.

Consider Figure 3-3 having

$$X_1, X_2, X_3, X_4 \ldots X_n$$

as inputs.

The corresponding weights are

$$W_1, W_2, W_3, W_4 \ldots W_n.$$

The product of the inputs and the weights is summated, and a bias is added to the result. That is,

$$U_i = \sum W_i X_i + b$$

This result U_i passes through a non-linear activation function "f" resulting in V_i:

$$V_i = f(U_i)$$

One of the most common activations used in the case of Neural Networks is the sigmoid function, which is given by the following equation:

$$f(x) = \frac{1}{1+e^{-x}}$$

The so-obtained value (V_i) is passed through a threshold (in the case of classification). Note that the same model can be used for regression, in which case thresholding is not done. In this model, the weights and bias are initialized to random numbers and then updated in each iteration.

The formal algorithm of SLP is as follows:

1. Initialize the weights (W) and biases (b) to random numbers between 0 and 1.

2. For each input sample X_i, calculate the net input as U_i by taking the dot product of the input features and the weights and adding the bias, that is,

$$\Sigma W_i X_i + b.$$

3. Compute the activation value V_i or $\hat{y}$ by passing the net input U_i through a nonlinear activation function:

$$\hat{y} = f(\Sigma W_i X_i + b)$$

4. Update the weights and bias using the formula

$$W = W - \alpha\, f(1-f)(\hat{y}-y)X_i,$$

$$b = b - \alpha\, f(1-f)(\hat{y}-y)$$

5. Repeat steps 2 to 4 till convergence is reached or the number of iterations becomes equal to the number of samples available.

Figure 3-3 shows the SLP model.

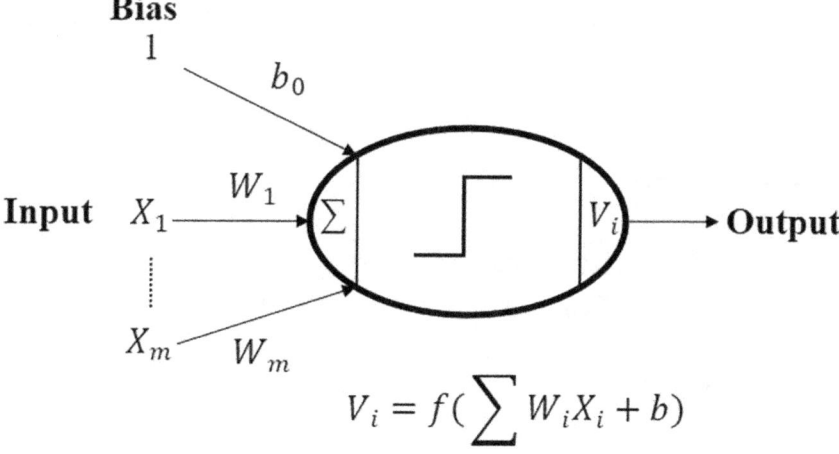

Figure 3-3. Single-Layer Perceptron model

Implementation of a SLP

The following code implements SLP (Listing 3-1). The code uses the first 100 samples of the popular IRIS dataset having four features. Each sample in the first 100 samples belongs to one of the two classes (binary classification) Setosa and Versicolor.

Let the weights corresponding to the four inputs be $[w_1, w_2, w_3, w_4]$ and the bias be b. The input to the output neuron would be the dot product of the weights and inputs, followed by the addition of bias. This sum is then passed through the activation function. The so-obtained output is compared with the expected output, and the squared error is evaluated.

The weights of the model are then updated in the following iterations, till there is no further change in the weights or for a pre-decided number of iterations.

The so-obtained weights are then used for predicting the class of an unseen sample for evaluating the given model.

Note that the hyperparameter α (learning rate) affects the performance of the model. The output shown in Figure 3-4 shows the variation of performance with α. Chapter 5 discusses the hyperparameters of Neural Networks in detail.

Listing 3-1. Implementing SLP from scratch to classify the IRIS dataset (first two classes)

Code:
```
#1. Importing Libraries
import numpy as np
from sklearn.datasets import load_iris
from sklearn.model_selection import train_test_split
from matplotlib import pyplot as plt
#2. Loading the Dataset
Data=load_iris()
X=Data.data
y=Data.target
print(X.shape)
print(y.shape)
```

#3. Selecting the first 100 samples
```
X=X[:100]
y=y[:100]
print(X.shape)
print(y.shape)
```
#4. Initializing weights and bias
```
def init_(X):
  w=np.random.random(X.shape[1])
  b=np.random.random()
  return w, b
```
#5. Min-Max Normalization
```
def normalise(X):
  max=np.max(X, axis=0)
  min=np.min(X, axis=0)
  return ((X-min)/(max-min))
```
#6. Sigmoid Activation Function
```
def f(x):
  return ((1)/(1+np.exp(-1*x)))
```
#7. Training the Model
```
def train(X_train, y_train, w, b, alpha):
  for i in range(X_train.shape[0]):
    x=X_train[i,:]
    u=np.sum(x*w)+b
    v=f(u)
    if v>0.5:
       y_pred=1
    else:
       y_pred=0
    w=w-alpha*(y_pred-y_train[i])*x
    b=b-alpha*(y_pred-y_train[i])
  return w, b
```

#8. Testing the model
```
def test(X_test, y_test, w, b):
        tp=0
        fp=0
```

CHAPTER 3 NEURAL NETWORKS

```
            tn=0
            fn=0
            for i in range(X_test.shape[0]):
                x=X_test[i,:]
                u=np.sum(x*w)+b
                v=f(u)
                if v>0.5:
                    y_pred=1
                else:
                    y_pred=0
                if(y_pred==1 and y_test[i]==1):
                    tp+=1
                elif(y_pred==0 and y_test[i]==0):
                    tn+=1
                elif(y_pred==1 and y_test[i]==0):
                    fp+=1
                else:
                    fn+=1
        accuracy=((tp+tn)/(tp+tn+fp+fn))*100
        return accuracy
```

#9. Driver Code
```
X_Norm=normalise(X)
y_Norm=normalise(y)
w, b=init_(X_Norm)
X_train, X_test, y_train, y_test=train_test_split(X_Norm, y_Norm, 
test_size=0.3)
result=[]
alpha=np.linspace(0.0001,0.1,500)
for i in alpha:
        w, b=train(X_train, y_train, w, b, i)
        accuracy=test(X_test, y_test, w, b)
result.append(accuracy)
best=np.max(result)
index=np.argmax(result)
print(best, index)
```

```
print(alpha[index])
plt.plot(alpha, result)
```

Output:

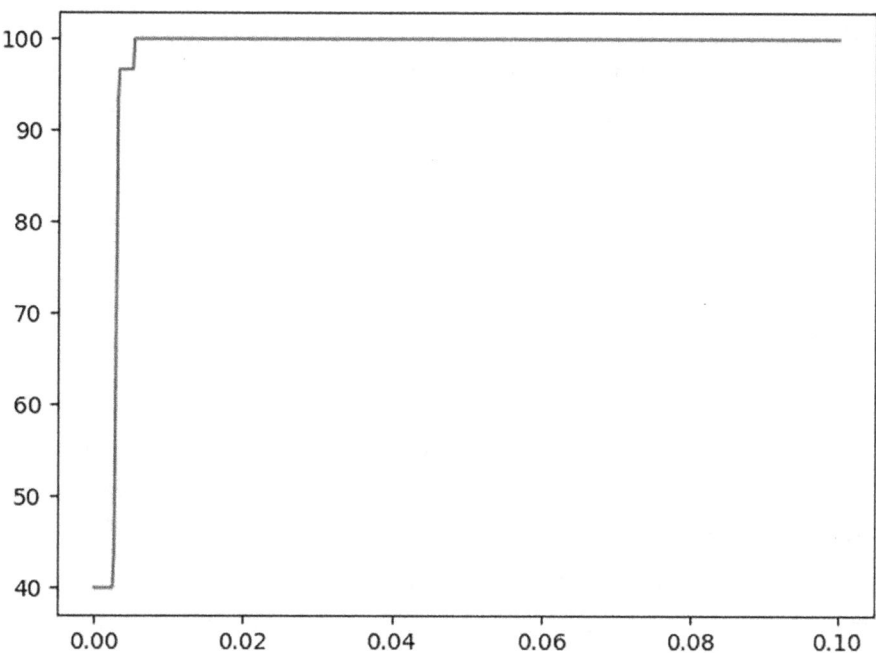

Figure 3-4. *Variation of performance with learning rate*

The following code (Listing 3-2) implements SLP using ***sklearn.linear_model.Perceptron*** on the Breast Cancer dataset containing 569 samples and 30 features. Table 3-1 shows the functions used for implementing the SLP.

Table 3-1. *sklearn Functions for Implementing Perceptron and Their Description*

Function	Description
perceptron = Perceptron ()	Initializes the classification algorithm.
perceptron.fit(X_train, y_train)	Fits or trains the model with the training set.
perceptron.predict(X_test)	Predicts the class for each sample in X_test.
accuracy_score(y_test, y_pred)	Calculates the accuracy of the model.

CHAPTER 3   NEURAL NETWORKS

Listing 3-2. Implementing SLP using the sklearn module to classify the Breast Cancer dataset

Code:
#1. Importing Libraries
```
import numpy as np
from sklearn.datasets import load_breast_cancer
from sklearn.model_selection import train_test_split
from sklearn.linear_model import Perceptron
from sklearn.metrics import accuracy_score
```
#2. Load the Breast Cancer dataset
```
breast_cancer = load_breast_cancer()
X = breast_cancer.data
y = breast_cancer.target
```
#3. Train Test Split
```
X_train, X_test, y_train, y_test = train_test_split(X, y, test_size=0.2, random_state=42)
```
#4. Fit the Model
```
perceptron = Perceptron(max_iter=1000, tol=1e-3, random_state=42)
perceptron.fit(X_train, y_train)
```
#5. Evaluate the Model
```
y_pred = perceptron.predict(X_test)
accuracy = accuracy_score(y_test, y_pred)
print(f"Accuracy: {accuracy}")
```

Having seen the implementation by scratch and using ***sklearn***, let's move to the implementation of SLP using ***Keras***. The following code (Listing 3-3) implements the SLP using ***Keras*** on the Breast Cancer dataset having 30 features and 569 samples. A sequential model is created with a dense layer having a single neuron. The model is compiled with a stochastic gradient descent (SGD) optimizer, binary cross-entropy (loss function), and accuracy (metric). It is trained on training data for 50 epochs with a batch size of 32. The model's loss and accuracy on training and test sets are shown in Figure 3-5.

Listing 3-3. Implementing SLP using the Keras module to classify the Breast Cancer dataset

Code:
#1. The libraries *keras.models* and *keras.layers* are imported to design a sequential model having dense layers. We need to import the *train_test_split* from *sklearn.model_selection* module for splitting the data into train and test sets.
```
import numpy as np
import pandas as pd
from sklearn.model_selection import train_test_split
from keras.models import Sequential
from keras.layers import Dense
from sklearn.datasets import load_breast_cancer
```
#2. The breast cancer dataset is loaded using *load_breast_cancer* function.
```
data = load_breast_cancer()
X = data.data
y = data.target
print(X.shape, y.shape)
```
#3. Train Test Split
```
X_train, X_test, y_train, y_test = train_test_split(X, y, test_size = 0.3)
print(X_train.shape, X_test.shape, y_train.shape, y_test.shape)
```
#4. The model having an input layer and a dense layer of single neuron with sigmoid activation is created. The model is compiled with an 'sgd' optimizer, binary cross entropy loss (binary classification), and accuracy metric. The model is trained over 50 epochs with the training set.
```
model_1 = Sequential()
model_1.add(Dense(units=1, input_dim= X.shape[1], activation='sigmoid'))
model_1.compile(optimizer='sgd', loss='binary_crossentropy', metrics=['accuracy'])
history = model_1.fit(X_train, y_train, epochs=50, batch_size=32, validation_data=(X_test, y_test))
loss, accuracy = model_1.evaluate(X_train, y_train)
print(f"Loss: {loss}, Accuracy: {accuracy}")
```

#5. Note that after compiling the model the output was saved in a variable called history. This is a dictionary from which training and validation accuracy and loss are plotted.

```
import matplotlib.pyplot as plt
train_loss = history.history['loss']
val_loss = history.history['val_loss']
train_acc = history.history['accuracy']
val_acc = history.history['val_accuracy']
plt.figure(figsize=(10, 5))
plt.subplot(1, 2, 1)
plt.plot(train_loss, label='Training Loss')
plt.plot(val_loss, label='Validation Loss')
plt.xlabel('Epoch')
plt.ylabel('Loss')
plt.title('Training and Validation Loss')
plt.legend()
plt.subplot(1, 2, 2)
plt.plot(train_acc, label='Training Accuracy')
plt.plot(val_acc, label='Validation Accuracy')
plt.xlabel('Epoch')
plt.ylabel('Accuracy')
plt.title('Training and Validation Accuracy')
plt.legend()
plt.tight_layout()
plt.show()
```

Output:

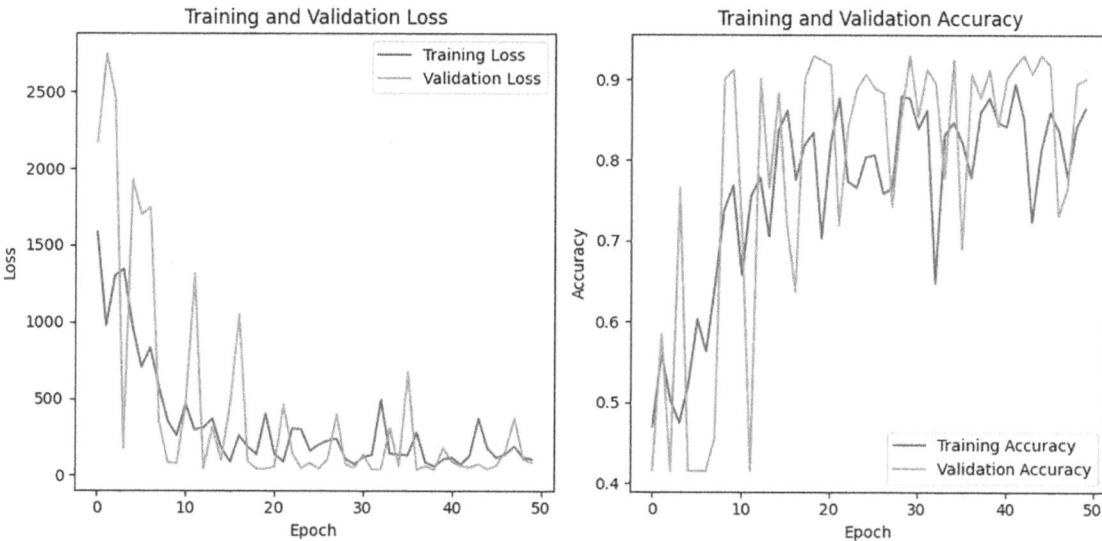

Figure 3-5. *Variation of loss and performance with the number of epochs (Listing 3-3)*

It can be seen that the loss decreases (in general) with the number of epochs and the performance improves. Let's use SLP for classifying a slightly complex dataset.

The following code (Listing 3-4) implements the SLP using **Keras** on **Myocardial Infarction Complications** having 1700 samples and 109 features after preprocessing the data. The architecture and the training process are the same as the previous model. The model's loss and accuracy on training and test sets are shown in Figure 3-6.

Listing 3-4. Implementing SLP using the Keras module to classify the Myocardial Infarction Complications dataset

Code:

#1. The `ucimlrep` is installed and fetched to import the `myocardial_infarction_complications` dataset.
```
!pip install ucimlrepo
from ucimlrepo import fetch_ucirepo
myocardial_infarction_complications= fetch_ucirepo(id=579)
X = myocardial_infarction_complications.data.features
y = myocardial_infarction_complications.data.targets
y = y['ZSN']
```

CHAPTER 3 NEURAL NETWORKS

#2. The NaNs are calculated for each feature and droppedthose having greater than threshold.

```
nan_count_per_column = X.isnull().sum()
print(nan_count_per_column)
threshold = len(X)*0.3
df = X.dropna(axis=1, thresh=threshold)
print(df)
```

#3. From *sklearn.impute* module the KNN imputer is imported to impute the remaining NaN values in the dataset.

```
import pandas as pd
from sklearn.impute import KNNImputer
imputer = KNNImputer()
df_imputed = pd.DataFrame(imputer.fit_transform(df), columns=df.columns)
print(df_imputed)
X = df_imputed
print(X.shape, y.shape)
```

#3. From *sklearn.model_selection* module the *train_test_split*function is imported to split the data into train and test.

```
from sklearn.model_selection import train_test_split
X_train, X_test, y_train, y_test = train_test_split(X, y, test_size = 0.3)
print(X_train.shape, X_test.shape, y_train.shape, y_test.shape)
```

#4. The model having an input layer and a dense layer of single neuron with sigmoid activation is created. The model is complied with an 'sgd' optimizer, binary cross entropy loss (binary classification), and accuracy metric. The model is trained over 50 epochs with the training set.

```
import numpy as np
from keras.models import Sequential
from keras.layers import Dense
model_1 = Sequential()
model_1.add(Dense(units=1, input_dim= X.shape[1], activation='sigmoid'))
model_1.compile(optimizer='sgd', loss='binary_crossentropy',
metrics=['accuracy'])
history = model_1.fit(X_train, y_train, epochs=50, batch_size=32,
validation_data=(X_test, y_test))
```

```python
loss, accuracy = model_1.evaluate(X_train, y_train)
print(f"Loss: {loss}, Accuracy: {accuracy}")
```
#5. Note that after compiling the model the output was saved in a variable called history. This is a dictionary from which training and validation accuracy and loss are plotted.
```python
import matplotlib.pyplot as plt
train_loss = history.history['loss']
val_loss = history.history['val_loss']
train_acc = history.history['accuracy']
val_acc = history.history['val_accuracy']
plt.figure(figsize=(10, 5))
plt.subplot(1, 2, 1)
plt.plot(train_loss, label='Training Loss')
plt.plot(val_loss, label='Validation Loss')
plt.xlabel('Epoch')
plt.ylabel('Loss')
plt.title('Training and Validation Loss')
plt.legend()
plt.subplot(1, 2, 2)
plt.plot(train_acc, label='Training Accuracy')
plt.plot(val_acc, label='Validation Accuracy')
plt.xlabel('Epoch')
plt.ylabel('Accuracy')
plt.title('Training and Validation Accuracy')
plt.legend()
plt.tight_layout()
plt.show()
```

CHAPTER 3 NEURAL NETWORKS

Output:

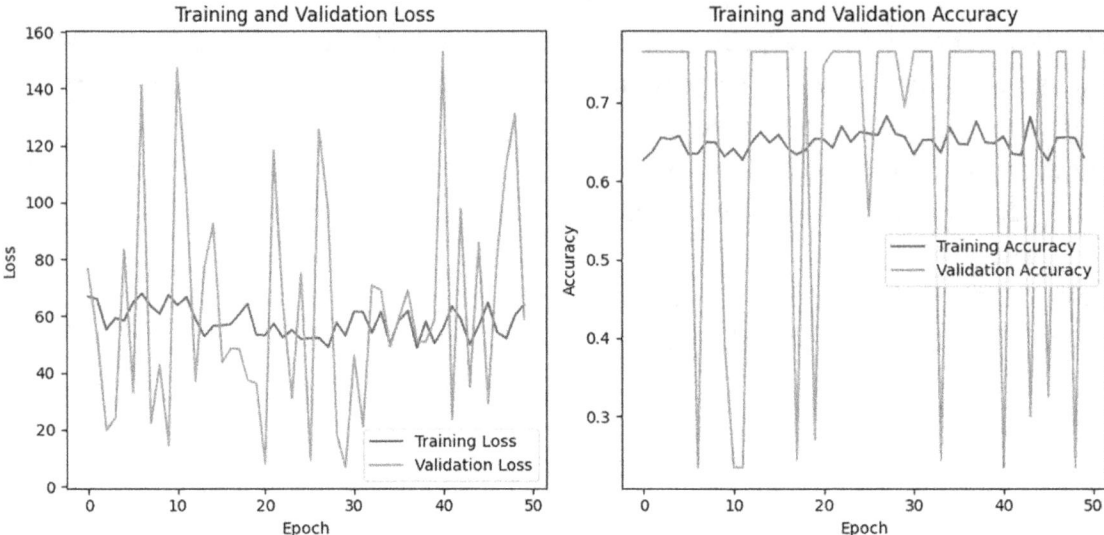

Figure 3-6. *Variation of loss and performance with the number of epochs (Listing 3-4)*

The results of the above models are summarized in Table 3-2. In this case, the results are not perfect since this data is not linearly separable.

Table 3-2. *Results of SLP with Two Different Datasets*

SLP No.	Dataset	Model	Accuracy	Loss
1.	Breast Cancer	SLP model_1 (with a single neuron in the output layer)	0.907	84.899
2.	Myocardial Infarction Complications	SLP model_1 (with a single neuron in the output layer)	0.7697	57.6161

As stated earlier, SLP can classify linearly separable inputs. However, when the input is not linearly separable, SLP might not work well. Let us have a look at a famous problem that cannot be solved using SLP: the XOR problem.

XOR Problem

Assume that you have two input variables (binary), to be segregated into two classes as shown in Table 3-3.

Table 3-3. XOR Table

A	B	Y
0	0	0
0	1	1
1	0	1
1	1	0

Figure 3-7 shows the value of (A, B) and the corresponding values of Y. Y=0 is shown using circles and Y=1 using triangles. Note that the circles and triangles cannot be segregated using a single line.

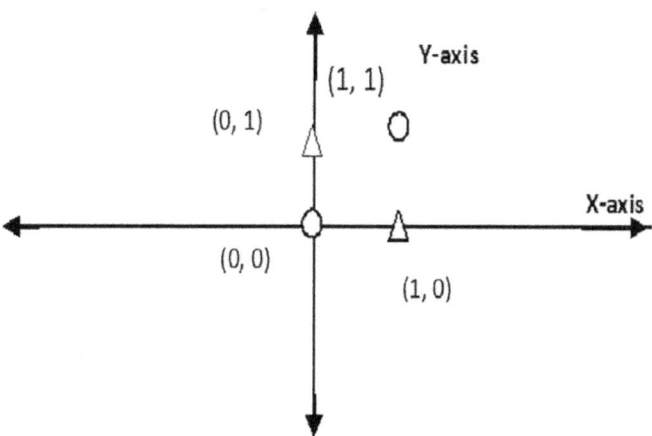

Figure 3-7. XOR problem

XOR Problem The XOR problem requires a classifier to be created that can classify the outputs of the XOR function treating the inputs of this function as two dimensions.

Since the data is not linearly separable, we cannot use SLP to classify the data. The Multi-layer Perceptron, discussed later in the chapter, will help us in solving the XOR problem.

CHAPTER 3 NEURAL NETWORKS

Activation functions play an important role in the recital of the model. Before proceeding further let us have a look at some of the most famous activation functions.

Activation Functions

This section presents a brief overview of the activation functions used in Neural Networks. The formula, range, derivative, and problems associated with each activation function are summarized in this section.

1. Sigmoid

The sigmoid activation function can be represented using the following equation:

$$f(x) = \frac{1}{1+e^{-x}}.$$

The graph of this function is shown in Figure 3-8.

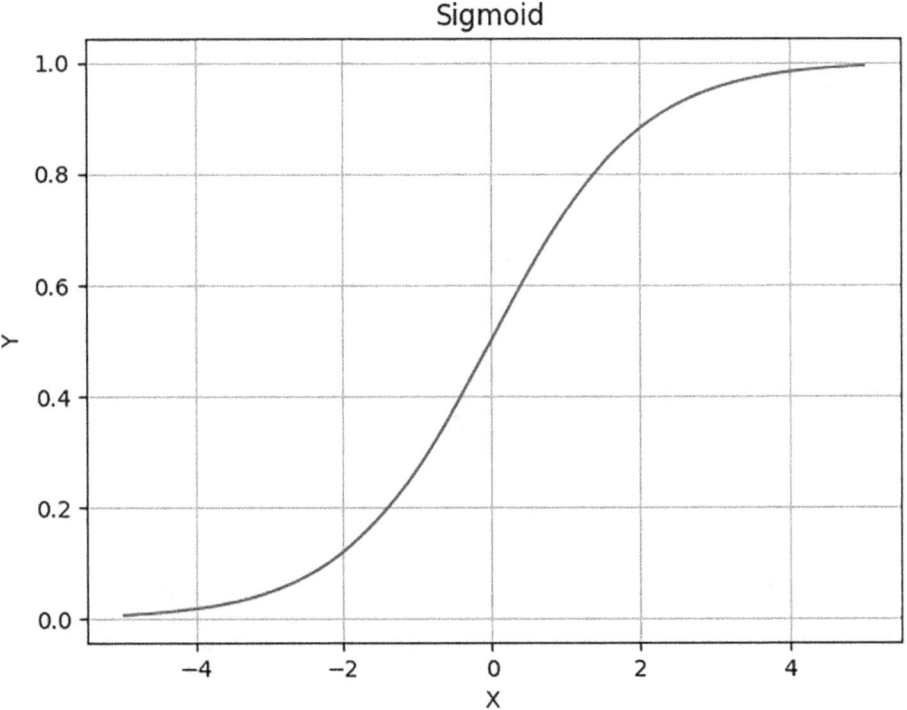

Figure 3-8. *Sigmoid activation function graph*

It is a smooth and differentiable function whose output range is between 0 and 1, which makes it suitable for representing output that depicts probabilities. However, it suffers from a vanishing gradient problem as discussed in Multi-layer Perceptron. This problem can slow down the learning process in the case of deeper networks; hence, newer activation functions like ReLU were later proposed by the researchers.

2. Tanh

The tanh activation function can be represented using the following equation:

$$f(x) = \frac{e^x - e^{-x}}{e^x + e^{-x}}.$$

The graph of this function is shown in Figure 3-9.

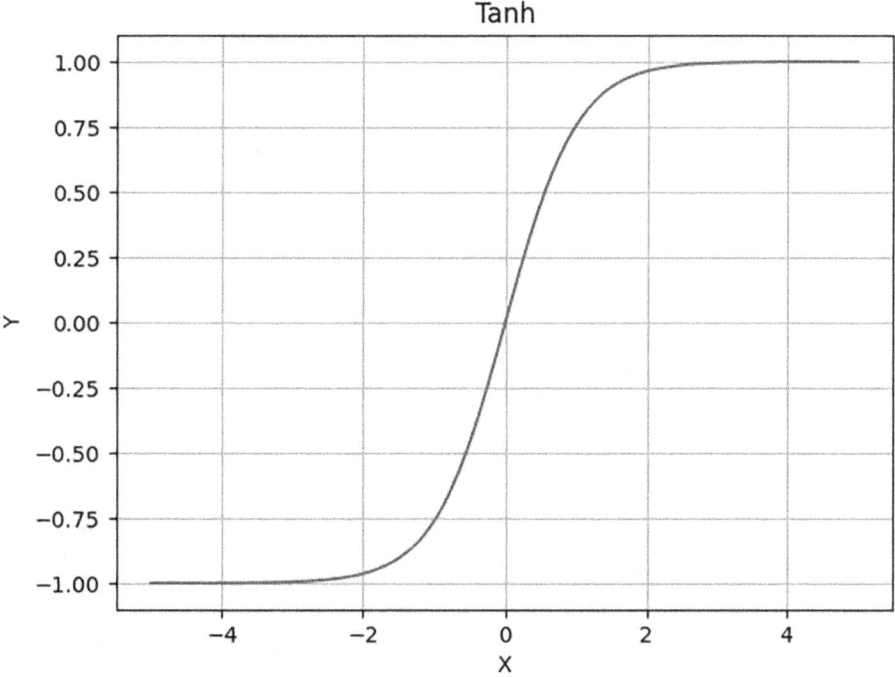

Figure 3-9. *Tanh activation function graph*

It is a smooth and differentiable function having an output range between -1 and 1. This is zero-centered as against the sigmoid function. This function also suffers from the vanishing gradient problem.

CHAPTER 3 NEURAL NETWORKS

3. Rectified Linear Unit (ReLU)

The ReLU activation function can be represented using the following equation:

$$f(x) = max(0,x).$$

The graph of this function is shown in Figure 3-10.

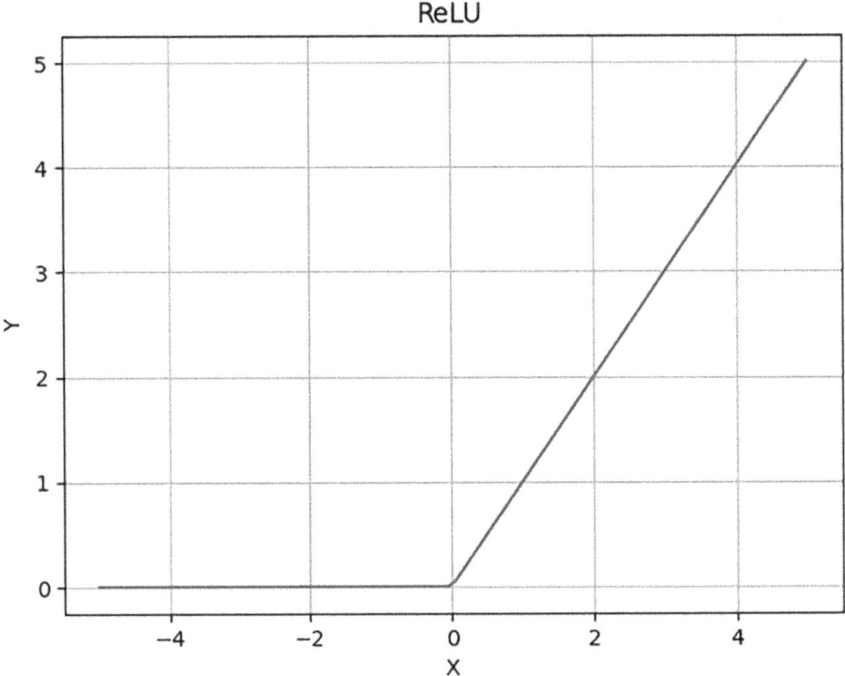

Figure 3-10. *ReLU activation function graph*

This is one of the most computationally efficient activation functions whose output range is between 0 and infinity, and it handles the problem of vanishing gradient gracefully. One of the problems faced by using these functions is that **if the input is negative, then the output becomes zero**. In addition to this, if the output of these functions is not bounded, then it results in a problem called **exploding gradient**.

4. Softmax

In the case of multiclass classification problems, softmax is considered one of the best activation functions. In softmax the output of a particular neuron in the output layer is given by the following formula:

$$f(x_i) = \frac{e^{x_i}}{\sum_j e^{x_j}}.$$

The graph of this function is shown in Figure 3-11.

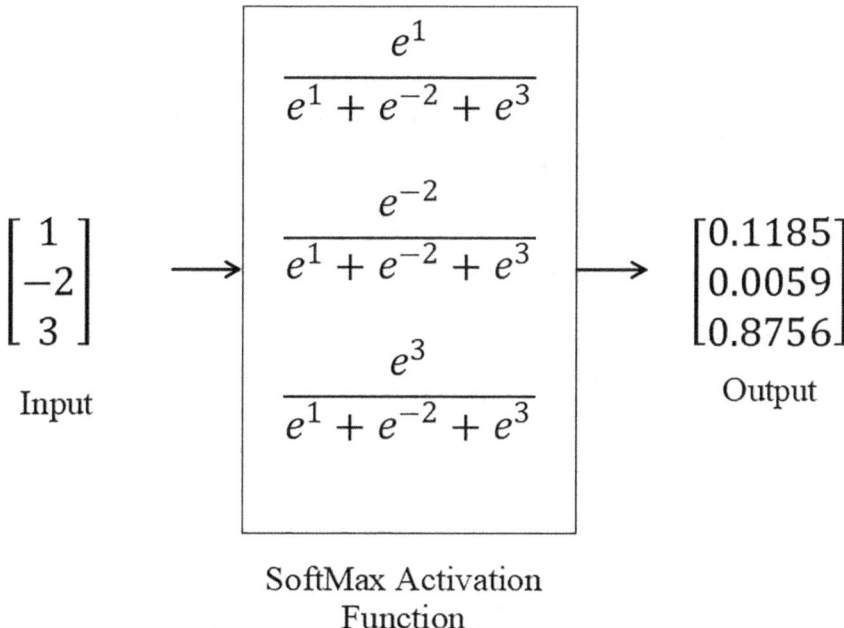

Figure 3-11. Softmax activation function graph

Note that the output range of each neuron is between 0 and 1 and these outputs may be considered as the probabilities whose sum is 1. So the neuron having the highest probability may be selected as the output.

CHAPTER 3 NEURAL NETWORKS

Multi-layer Perceptron

We have already seen that a SLP forms a linear combination of input features and gives it as an argument to a nonlinear activation function. Now, imagine that various such combinations of input features are created in a layer, and they act as input to the next layer, thus creating a **hierarchy of features**. The Multi-layer Perceptron does create a hierarchical feature representation and can handle nonlinearly separable data. Let's begin with solving the XOR problem (nonlinearly separable data) using MLP.

Solving the XOR Problem Using Multi-layer Perceptron

Let us consider an **XOR** gate. We have already seen that it cannot be implemented by a SLP. However, **AND** and **OR** gates can be implemented using SLP. We have also seen earlier that a **NAND** gate can be created in the same way as an **AND** gate with negated inputs. Now, consider a network having two inputs X_1 and X_2. You can easily create a SLP for the implementation of the **NAND** gate and **OR** gate as shown in Figure 3-12.

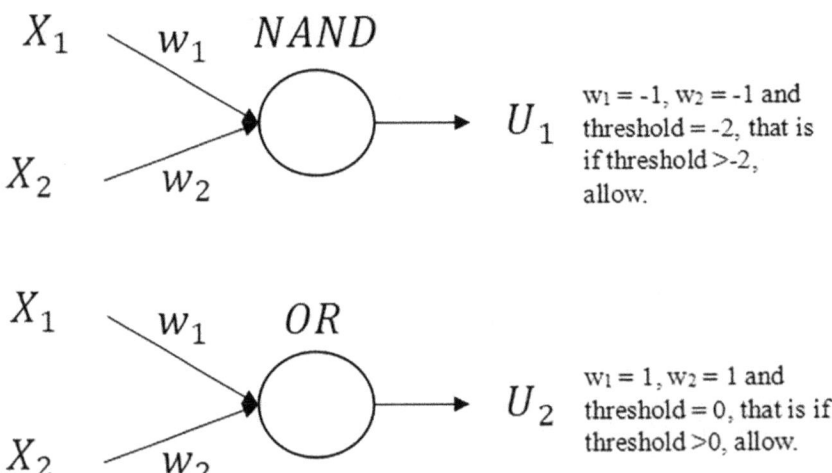

Figure 3-12. Implementing NAND and OR gates using SLP

To construct an **XOR** gate, the output of the above networks acts as an input to a neuron in the next layer, which implements the **AND** gate shown in Figure 3-13.

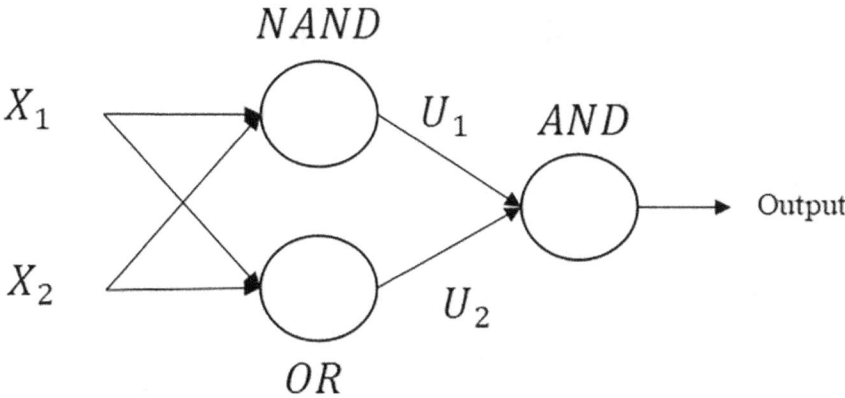

Figure 3-13. *Implementing an XOR gate using **NAND**, **OR**, and **AND** gates*

Let us see why the above construction is mathematically correct. As we understand XOR can be represented by the following Boolean expression:

$$Y = A\bar{B} + \bar{A}B$$

NAND can be represented as

$$Y \equiv \overline{A.B}$$

which can be written as follows (applying De Morgan's Law).

$$Y = \bar{A} + \bar{B}$$

Now, multiplying $A + B$ with Y, we get

$$Z = (\bar{A} + \bar{B})(A + B)$$

$$Z = \bar{A}A + \bar{A}B + \bar{B}A + \bar{B}B$$

$$Z = \bar{A}B + A\bar{B}$$

which is the same as XOR. Therefore, it can be concluded that XOR can be perceived as **AND** of **NAND** and **OR**. Also, **NAND** and **OR** can be implemented by SLP. It implies that XOR can be recreated using two layers of SLP and can classify nonlinearly separable data.

Tip The multi-layer Neural Network can classify nonlinearly separable data.

CHAPTER 3　NEURAL NETWORKS

Architecture of MLP and Forward Pass

A Multi-layer Perceptron has an input layer, an output layer, and at least one hidden layer. Figure 3-14 shows the architecture of a MLP having **n** inputs and a single output. Assume that there is only one hidden layer having **p** neurons.

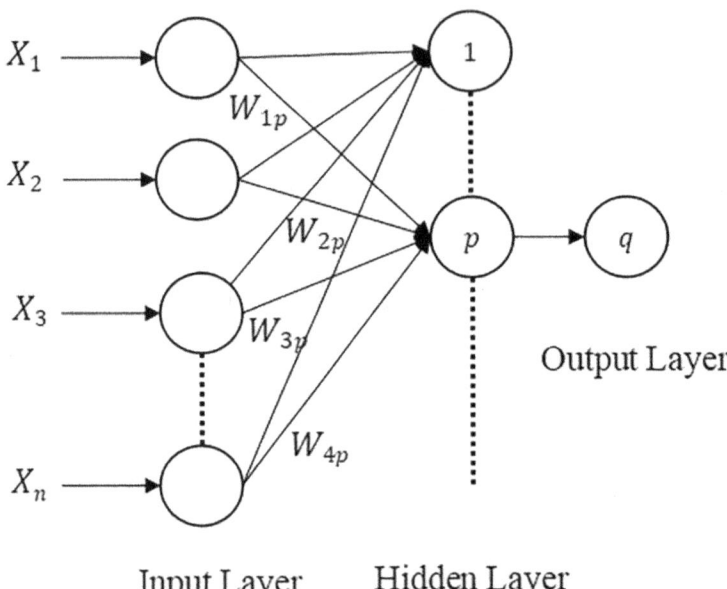

Figure 3-14. Multi-layer Perceptron having n neurons in the input layer and a single neuron in the output layer

Let the inputs be $X_1, X_2, X_3, X_4....X_n$ and the weights between the first and the second layer are W_{ij}. Consider a particular neuron, say **p**, in the hidden layer.

At the pth neuron in the hidden layer, the input features, multiplied by the corresponding weights, added with the bias become the input to the activation function:

$$U_p = \sum X_i W_{ip} + b_p$$

The output of the pth neuron can be represented as

$$V_p = f(U_p)$$

where f is the activation function. Likewise, the input to all the neurons in the hidden layer can be calculated.

CHAPTER 3 NEURAL NETWORKS

Now consider a neuron (q) at the output layer. The output of this neuron can be calculated as follows:

$$U_q = \sum V_p W_{pq} + b_q$$

$$V_q = f(U_q)$$

At each layer, we process the input and calculate the output, which becomes the input to the next layer. This network would henceforth be referred to as **Feed-Forward Network**.

As an example, consider a network to classify the standard IRIS dataset having four features. From this dataset consider the first 100 samples, having two classes: Setosa and Versicolor. Let us develop a network to classify this dataset. The network has four neurons in the input layer, two neurons in the hidden layer (how?), and one neuron in the output layer, as shown in Figure 3-15. The weights from the input to the first hidden layer have 1 as superscript and those from hidden to the output have 2 as superscript.

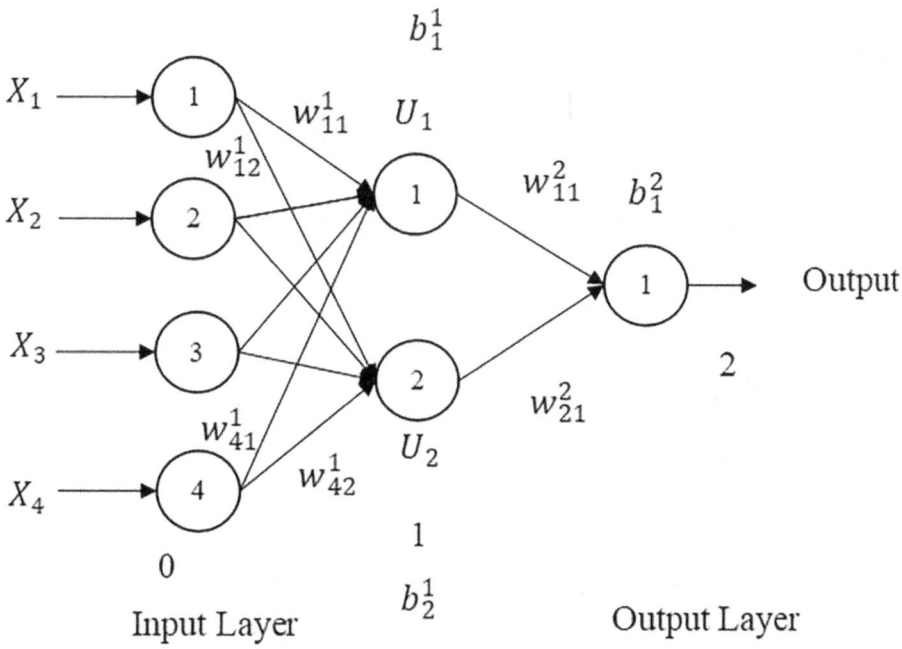

Figure 3-15. *Architecture of the network having four neurons in the input layer and two neurons in the hidden layer and a single neuron in the output layer*

Let us now calculate the values of the outputs of each layer in the feed-forward pass, assuming that initial weights and bias are given.

83

CHAPTER 3 NEURAL NETWORKS

Feed-Forward

$$U_1 = \sum_i X_i w_{i1}^1 + b_1^1$$

$$V_1 = f\left(\sum_i X_i w_{i1}^1 + b_1^1\right) = f\left((X_1 \times w_{11}^1) + (X_2 \times w_{21}^1) + (X_3 \times w_{31}^1) + (X_4 \times w_{41}^1) + b_1^1\right)$$

$$U_2 = \sum_i X_i w_{i2}^1 + b_2^1$$

$$V_2 = f\left(\sum_i X_i w_{i2}^1 + b_2^1\right) = f\left((X_1 \times w_{12}^1) + (X_2 \times w_{22}^1) + (X_3 \times w_{32}^1) + (X_4 \times w_{42}^1) + b_2^1\right)$$

Now the output of the network can be represented as

$$O_i = f\left(\sum_j V_j w_{j1}^2 + b_1^2\right) = f\left((V_1 \times w_{11}^2) + (V_2 \times w_{21}^2) + b_1^2\right)$$

The value obtained using the above calculations (forward pass) is the output of the network.

The output so obtained is then compared with the expected output, and error is calculated. This is followed by updating weights between the output and the hidden layer; the weights between the input and the hidden layer are then updated. The process is examined in the following section.

Gradient Descent

In a conventional Machine Learning pipeline, you generally preprocess the given data, extract features out of it, select the relevant features, make predictions, and design a loss function that minimizes the difference between the expected and the predicted value. In each iteration, the model tries to minimize this loss. To accomplish this task, one of the methods that is commonly employed is the gradient descent method. To understand the method, let us consider a SLP in which the weights and bias are initialized to random values. These parameters are multiplied with input features to give a linear combination that passes through a nonlinear activation function to generate a predicted value. In the case of classification, thresholding is done after this step.

Let the predicted value be $\hat{y} = f(W^T X + b)$ and the loss function be $\frac{1}{2}(\hat{y}_i - y_i)^2$, that is, the squared difference between the expected and the predicted value (1/2 is multiplied for mathematical convenience). The gradient of loss with respect to weight can be calculated as follows:

$$\frac{\partial L}{\partial W} = \frac{\partial \left(\left(\frac{1}{2}(\hat{y}_i - y_i)^2 \right) \right)}{\partial W}$$

$$\frac{\partial L}{\partial W} = \frac{\partial \left(\frac{1}{2}\left(f(W^T X + b) - y_i \right)^2 \right)}{\partial W}$$

$$\frac{\partial L}{\partial W} = \left(f(W^T X + b) - y_i \right) \times f(1-f) \times X$$

The weights are then updated using the following formula:

$$W_{new} = W_{old} - \propto \frac{\partial L}{\partial W}$$

Here, W_{old} is the value of weight in the previous iteration (some random value), and $\frac{\partial L}{\partial W}$ is calculated in the previous step. $\propto$ is the learning rate that determines the step size. If the value of $\propto$ is very small, then it will take a large amount of time to reach the optimal value. On the other hand, if the value of $\propto$ is very large, then it might skip the local minima. The formula for updating the value of bias is as follows:

$$b_{new} = b_{old} - \propto (\hat{y}_i - y_i) f(1-f).$$

The above procedure can be used to find the weights in each iteration for a single-layer Neural Network. However, for multiple layers updating the weights becomes problematic as explained earlier. For updating weights in a MLP, first start with the outermost layer. Update the weights using the above algorithm. Once we have updated the weights, we move backward and update the weights of the second last layer using the backpropagation algorithm.

CHAPTER 3 NEURAL NETWORKS

Backpropagation

Once we calculate the squared error by taking the square of the difference between the expected and the obtained value, we then proceed to update the weights of the network. To do so, we first update the weights between the output and the hidden layer using the formula obtained using gradient descent in the previous section. This is followed by updating the weights of the hidden layer using the backpropagation algorithm:

$$W_{ij}^k = W_{ij}^k - \eta \delta_j^k O_i^{k-1}$$

$$\delta_j^k = O_j^k \left(1 - O_j^k\right)\left(O_j^k - t_j\right)$$

Let's have a look at the backpropagation algorithm for learning the weights of the hidden layer.

Backpropagation Algorithm

1. Initialize the weights and biases for each layer with small random values.

2. For each layer (forward pass)

   a. Calculate the weighted sum of inputs for each neuron: $\sum X_i W_{ij} + b$.

   b. Apply the activation function $f(\sum X_i W_{ij} + b)$ to generate the output of that layer.

3. Calculate the error at the output layer: $\frac{1}{2}(\hat{y}_i - y_i)^2$

4. Calculate the gradient of loss for weights: $\dfrac{\partial L}{\partial W} = \dfrac{\partial \left(\left(\frac{1}{2}(\hat{y}_i - y_i)^2\right)\right)}{\partial W}$

5. Update the weights of the last layer using the computed gradient and with a learning rate ∝: $W_{new} = W_{old} - \propto \dfrac{\partial L}{\partial W}$

6. Update the weights of the hidden layer using the following equation.

For any hidden layer weight: W_{ij}^k

$$W_{ij}^k = W_{ij}^k - \eta \delta_i^k O_j^{k-1}$$

where

$$\delta_i^k = O_i^k\left(1-O_i^k\right)\sum_{j=1}^{M_{k+1}}\partial_j^{k+1}W_{ij}^{k+1}$$

$$\partial_j^{k+1} = O_j^{k+1}\left(1-O_j^{k+1}\right)\left(O_j^{k+1}-t_j\right)$$

7. Repeat the forward and backward pass for a predefined number of epochs or until convergence.

Implementation

The MLPs must contain at least one hidden layer. However, they can have multiple hidden layers as well. Note that the

- Number of hidden layers
- Number of neurons in the hidden layer
- Activation function
- Learning rate

are some of the important hyperparameters in the case of MLPs. One of the ways of determining these hyperparameters is empirical analysis. The topic is dealt with in Chapter 5.

To understand this, let us take an example. The example that follows classifies the wine dataset's first two classes having 13 features and 130 samples (Listing 3-5). Note that the implementation that follows uses the *sklearn.neural_network.MLPClassifier* function of the *sklearn* module. Two models have been created, one with the default number of neurons in the single hidden layer, that is, 100, and the other with only 3 neurons in the hidden layer.

Listing 3-5. Implementing MLP using the sklearn module to classify the first two classes of the wine dataset

Code:
```
#1. The Wine dataset is imported from sklearn.datasets using load_wine
function. The MLP classifier is imported from sklearn.neural_network
module. Additionally, train_test_split from sklearn.model_selection to
```

CHAPTER 3 NEURAL NETWORKS

split the data into train and test sets and *accuracy_score* from *sklearn.metrics* to evaluate the accuracy of the model have also been imported.

```
import numpy as np
from sklearn.datasets import load_wine
from sklearn.model_selection import train_test_split
from sklearn.neural_network import MLPClassifier
from sklearn.metrics import accuracy_score
```

#2. The wine dataset is loaded using *load_wine* function and the first two classes were selected.

```
wine = load_wine()
X = wine.data
y = wine.target
mask = y < 2
X = X[mask]
y = y[mask]
print(X.shape, y.shape)
```

#3. Train Test Split to split the data into train and test set

```
X_train, X_test, y_train, y_test = train_test_split(X, y, test_size=0.2, random_state=42)
```

#4. The Model 1: mlp_defaultisfittedwith the default parameters of MLP Classifier

```
mlp_default = MLPClassifier(random_state=42)
mlp_default.fit(X_train, y_train)
```

#5. The Model 2: mlp_custom is fittedwith the MLP Classifier having 3 neurons in a single hidden layer

```
mlp_custom = MLPClassifier(hidden_layer_sizes=(3,), random_state=42)
mlp_custom.fit(X_train, y_train)
```

#6. Using the predictions with default MLP, the accuracy score is calculated.

```
y_pred_default = mlp_default.predict(X_test)
accuracy_default = accuracy_score(y_test, y_pred_default)
print("Default MLP Accuracy: ", accuracy_default)
```

#6. Using the predictions with custom MLP, the accuracy score is calculated.

```
y_pred_custom = mlp_custom.predict(X_test)
accuracy_custom = accuracy_score(y_test, y_pred_custom)
print("Custom MLP (3 Neurons) Accuracy: ", accuracy_custom)
```

Output:
```
Default MLP Accuracy:  0.9230769230769231
Custom MLP (3 Neurons) Accuracy:  0.8076923076923077
```

Note that the model gives an accuracy of 92.3% with 100 neurons in the hidden layer and 80.76% with 3 neurons in the hidden layer. Neural Networks can have more than one hidden layer as well. The number of hidden layers and the number of neurons in each hidden layer can be found by various methods, one of which is empirical analysis. To understand this, consider the Breast Cancer dataset having 30 features and 569 samples. The following code implements two different models (Listings 3-6 and 3-7). The first model has a single hidden layer of 16 neurons, whereas the second model has two hidden layers of 16 and 8 neurons, respectively. By analyzing the results, one can infer that near-optimal performance can be obtained by multiple hidden layers or a single layer. However, the total number of parameters is different in both the cases. The following implementations also analyze the performance of the model by varying the learning rate and optimizers.

Listing 3-6. Implementing MLP using the Keras module to classify the Breast Cancer dataset

Code (single hidden layer of 16 neurons):
#1. The libraries *keras.models* and *keras.layers* are imported to design a sequential model having dense layers. We need to import the *train_test_split* from *sklearn.model_selection*module for splitting the data into train and test sets.
```
import numpy as np
import pandas as pd
from sklearn.model_selection import train_test_split
from keras.models import Sequential
from keras.layers import Dense
from sklearn.datasets import load_breast_cancer
```
#2. The breast cancer dataset is loaded using *load_breast_cancer* function.
```
data = load_breast_cancer()
X = data.data
y = data.target
print(X.shape, y.shape)
```

CHAPTER 3   NEURAL NETWORKS

#3. The *train_test_split* function is used to split the dataset into train and test set.
```
X_train, X_test, y_train, y_test = train_test_split(X, y, test_size = 0.3)
print(X_train.shape, X_test.shape, y_train.shape, y_test.shape)
```
#4. The model having an input layer and two dense layers of 16(hidden layer) and 1 (for output) neuron with sigmoid activation is created. The model is complied with'sgd' optimizer, binary cross entropy loss (binary classification), and accuracy metric. The model is trained over 50 epochs with the training set.
```
model_2 = Sequential()
model_2.add(Dense(units=16, input_dim= X.shape[1], activation='sigmoid'))
model_2.add(Dense(units=1, activation='sigmoid'))
model_2.compile(optimizer='sgd', loss='binary_crossentropy',
metrics=['accuracy'])
history = model_2.fit(X_train, y_train, epochs=50, batch_size=32,
validation_data=(X_test, y_test))
loss, accuracy = model_2.evaluate(X_test, y_test)
print(f"Loss: {loss}, Accuracy: {accuracy}")
```
#5. Note that after compiling the model the output was saved in a variable called history. This is a dictionary from which training and validation accuracy and loss are plotted.
```
import matplotlib.pyplot as plt
train_loss = history.history['loss']
val_loss = history.history['val_loss']
train_acc = history.history['accuracy']
val_acc = history.history['val_accuracy']
plt.figure(figsize=(10, 5))
plt.subplot(1, 2, 1)
plt.plot(train_loss, label='Training Loss')
plt.plot(val_loss, label='Validation Loss')
plt.xlabel('Epoch')
plt.ylabel('Loss')
plt.title('Training and Validation Loss')
plt.legend()
plt.subplot(1, 2, 2)
plt.plot(train_acc, label='Training Accuracy')
```

```python
plt.plot(val_acc, label='Validation Accuracy')
plt.xlabel('Epoch')
plt.ylabel('Accuracy')
plt.title('Training and Validation Accuracy')
plt.legend()
plt.tight_layout()
plt.show()
```
Output:

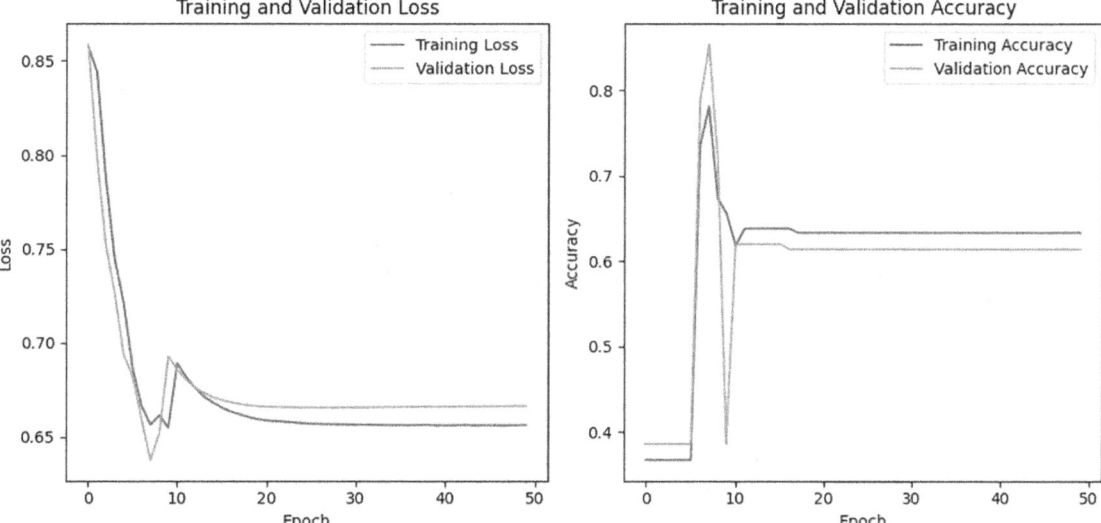

Figure 3-16. *Training and validation loss and accuracy variation with number of epochs*

Figure 3-16 (left) shows the variation of training and validation loss with the number of epochs, and Figure 3-16 (right) shows the variation of performance with the number of epochs for model 1.

The choice of different optimizers also affects the performance of a model and the variation in the loss. As you can see in Figures 3-17, 3-18, and 3-19, the variation of performance and loss with the learning rate with different optimizers gives different results. For this particular dataset and this model, the performance does not change with the learning rate in the case of stochastic gradient descent. However, the variation of loss is noticeable. In the case of RMSprop and Adam with the same model, the accuracy touches 90% on a learning rate of 10^{-1}. However, the variation of loss is stable.

CHAPTER 3   NEURAL NETWORKS

Stochastic Gradient Descent

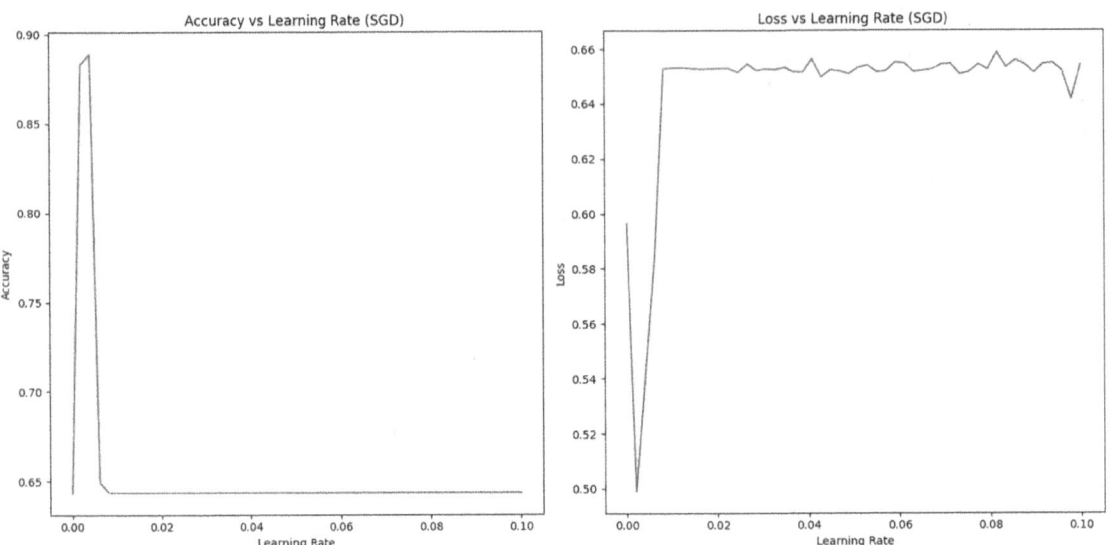

Figure 3-17. *Variation of loss and accuracy with learning rate for the stochastic gradient descent optimizer*

RMSprop

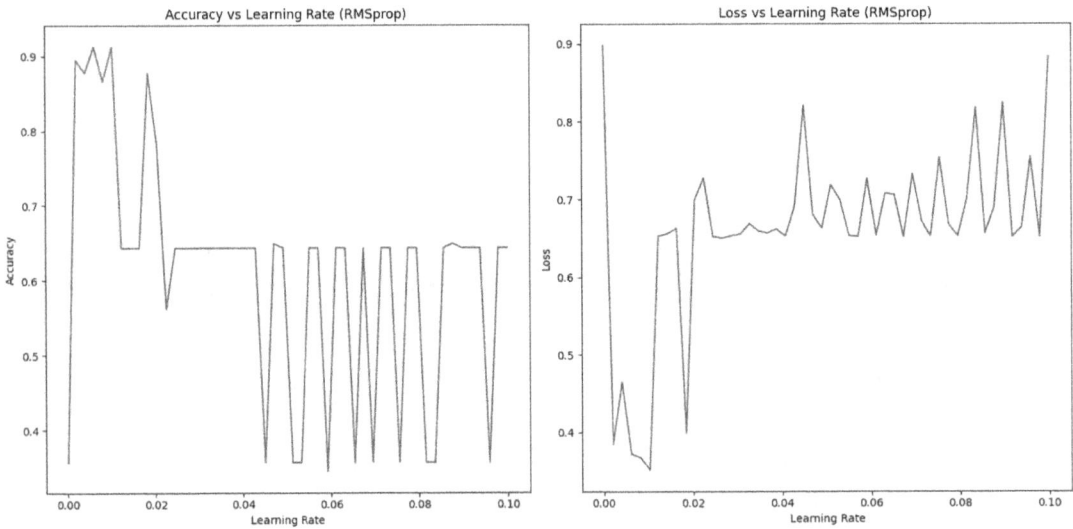

Figure 3-18. *Variation of loss and accuracy with learning rate for the RMSprop optimizer*

Adam

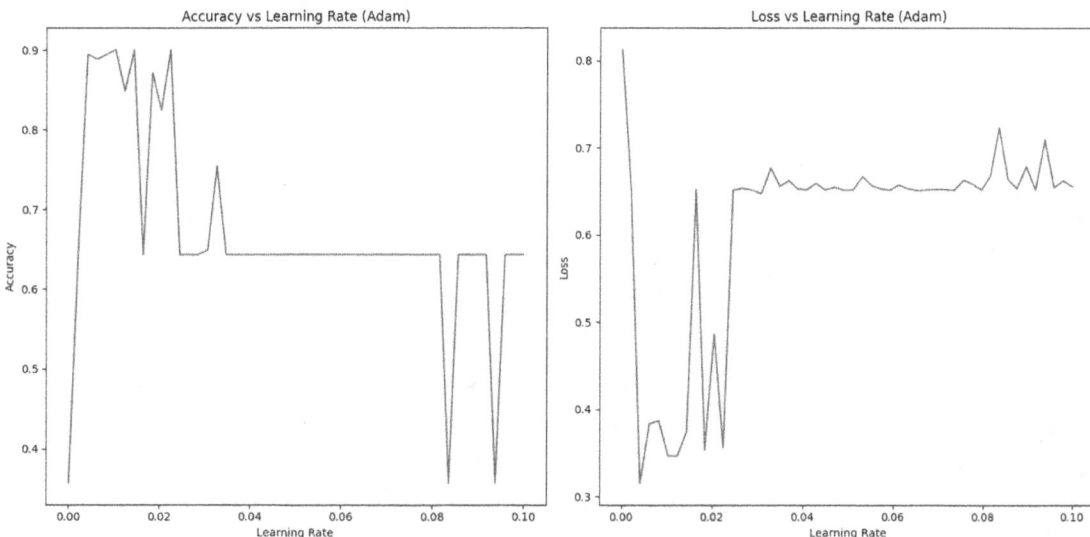

Figure 3-19. *Variation of loss and accuracy with learning rate for the Adam optimizer*

Now let us move to the implementation of Multi-layer Perceptron to classify the Breast Cancer dataset with two hidden layers using the **_Keras_** module (Listing 3-7).

Listing 3-7. Implementing MLP using the Keras module to classify the Breast Cancer dataset

Code (two hidden layers of 16 and 8 neurons):
#1. The libraries `keras.models` and `keras.layers` are imported to design a sequential model having dense layers. We need to import the `train_test_split` from `sklearn.model_selection` module for splitting the data into train and test sets.

```
import numpy as np
import pandas as pd
from sklearn.model_selection import train_test_split
from keras.models import Sequential
from keras.layers import Dense
from sklearn.datasets import load_breast_cancer
```

#2. The breast cancer dataset is loaded using the *load_breast_cancer* function.

```
data = load_breast_cancer()
X = data.data
y = data.target
print(X.shape, y.shape)
```

#3. The *train_test_split* function is used to split the dataset into train and test sets.

```
X_train, X_test, y_train, y_test = train_test_split(X, y, test_size = 0.3)
print(X_train.shape, X_test.shape, y_train.shape, y_test.shape)
```

#4. The model having an input layer with two dense layers of 16 and 8 (hidden layer) neurons followed by a dense layer of 1 (for output) neuron with sigmoid activation is created. The model is complied with 'sgd' optimizer, binary cross entropy loss (binary classification) and accuracy metric. The model is trained over 50 epochs with the training set.

```
model_3 = Sequential()
model_3.add(Dense(units=16, input_dim= X.shape[1], activation='sigmoid'))
model_3.add(Dense(units=8, activation='sigmoid'))
model_3.add(Dense(units=1, activation='sigmoid'))
model_3.compile(optimizer='sgd', loss='binary_crossentropy', metrics=['accuracy'])
history = model_3.fit(X_train, y_train, epochs=50, batch_size=32, validation_data=(X_test, y_test))
loss, accuracy = model_3.evaluate(X_test, y_test)
print(f"Loss: {loss}, Accuracy: {accuracy}")
```

#5. Note that after compiling the model the output was saved in a variable called history. This is a dictionary from which training and validation accuracy and loss are plotted.

```
import matplotlib.pyplot as plt
train_loss = history.history['loss']
val_loss = history.history['val_loss']
train_acc = history.history['accuracy']
val_acc = history.history['val_accuracy']
plt.figure(figsize=(10, 5))
plt.subplot(1, 2, 1)
```

```
plt.plot(train_loss, label='Training Loss')
plt.plot(val_loss, label='Validation Loss')
plt.xlabel('Epoch')
plt.ylabel('Loss')
plt.title('Training and Validation Loss')
plt.legend()
plt.subplot(1, 2, 2)
plt.plot(train_acc, label='Training Accuracy')
plt.plot(val_acc, label='Validation Accuracy')
plt.xlabel('Epoch')
plt.ylabel('Accuracy')
plt.title('Training and Validation Accuracy')
plt.legend()
plt.tight_layout()
plt.show()
```

Output:

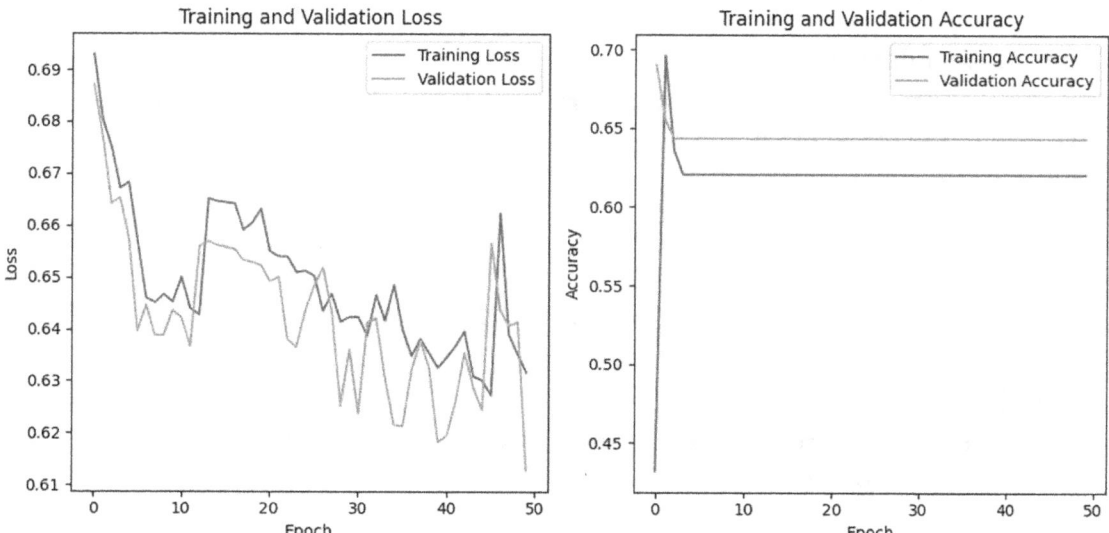

Figure 3-20. *Training and validation loss and accuracy variation with number of epochs*

Figure 3-20 (left) shows the variation of training and validation loss with the number of epochs, and Figure 3-20 (right) shows the variation of performance with the number of epochs for model 2.

CHAPTER 3 NEURAL NETWORKS

Here the model is trained through 50 epochs. Note that on increasing the number of epochs, the loss should decrease, whereas the performance should increase. The results are summarized in Table 3-4.

Table 3-4. *Results of the Above Models on the Breast Cancer Dataset*

MLP No.	Dataset	Model	Accuracy	Loss
1.	Breast Cancer	model_2 (single hidden layer with 16 neurons)	0.8538	0.664
2.		model_3 (two hidden layers with 16 and 8 neurons)	0.6901	0.6128

The above implementations are also used to classify the Myocardial Infarction Complications dataset, which is slightly complex and has 1700 samples and 109 features. The first implementation that follows contains a single hidden layer having 50 neurons (Listing 3-8). In the second implementation, the model contains two hidden layers having 25 and 12 neurons (Listing 3-9).

Listing 3-8. Implementing MLP with a single hidden layer of 50 neurons using the Keras module to classify the Myocardial Infarction Complications dataset

Code (single hidden layer of 50 neurons):
#1. The *ucimlrep* is installed and fetched to import the *myocardial_infarction_complications* dataset.
```
!pip install ucimlrepo
from ucimlrepo import fetch_ucirepo
myocardial_infarction_complications = fetch_ucirepo(id=579)
X = myocardial_infarction_complications.data.features
y = myocardial_infarction_complications.data.targets
y = y['ZSN']
```
#2. The NaNs are calculated for each feature and dropped those having greater than threshold.
```
nan_count_per_column = X.isnull().sum()
print(nan_count_per_column)
threshold = len(X)*0.3
df = X.dropna(axis=1, thresh=threshold)
print(df)
```

#3. From *sklearn.impute* module the KNN imputer is imported to impute the remaining NaN values in the dataset.

```
import pandas as pd
from sklearn.impute import KNNImputer
imputer = KNNImputer()
df_imputed = pd.DataFrame(imputer.fit_transform(df), columns=df.columns)
print(df_imputed)
X = df_imputed
print(X.shape, y.shape)
```

#4. From *sklearn.model_selection* module the *train_test_split* function is imported to split the data into train and test.

```
from sklearn.model_selection import train_test_split
X_train, X_test, y_train, y_test = train_test_split(X, y, test_size = 0.3)
print(X_train.shape, X_test.shape, y_train.shape, y_test.shape)
```

#5. The model has an input layer, a dense layer of 50 neurons(hidden layer), and a dense layer of 1 neuron (for output) with sigmoid activation is created. The model is complied with an 'sgd' optimizer, binary cross entropy loss (binary classification), and accuracy metric. The model is trained over 50 epochs with the training set.

```
model_2 = Sequential()
model_2.add(Dense(units=50, input_dim= X.shape[1], activation='sigmoid'))
model_2.add(Dense(units=1, activation='sigmoid'))
model_2.compile(optimizer='sgd', loss='binary_crossentropy',
metrics=['accuracy'])
history = model_2.fit(X_train, y_train, epochs=50, batch_size=32,
validation_data=(X_test, y_test))
loss, accuracy = model_2.evaluate(X_train, y_train)
print(f"Loss: {loss}, Accuracy: {accuracy}")
```

#6. Note that after compiling the model the output was saved in a variable called history. This is a dictionary from which training and validation accuracy and loss are plotted.

```
import matplotlib.pyplot as plt
train_loss = history.history['loss']
val_loss = history.history['val_loss']
train_acc = history.history['accuracy']
```

CHAPTER 3 NEURAL NETWORKS

```
val_acc = history.history['val_accuracy']
plt.figure(figsize=(10, 5))
plt.subplot(1, 2, 1)
plt.plot(train_loss, label='Training Loss')
plt.plot(val_loss, label='Validation Loss')
plt.xlabel('Epoch')
plt.ylabel('Loss')
plt.title('Training and Validation Loss')
plt.legend()
plt.subplot(1, 2, 2)
plt.plot(train_acc, label='Training Accuracy')
plt.plot(val_acc, label='Validation Accuracy')
plt.xlabel('Epoch')
plt.ylabel('Accuracy')
plt.title('Training and Validation Accuracy')
plt.legend()
plt.tight_layout()
plt.show()
```

Output:

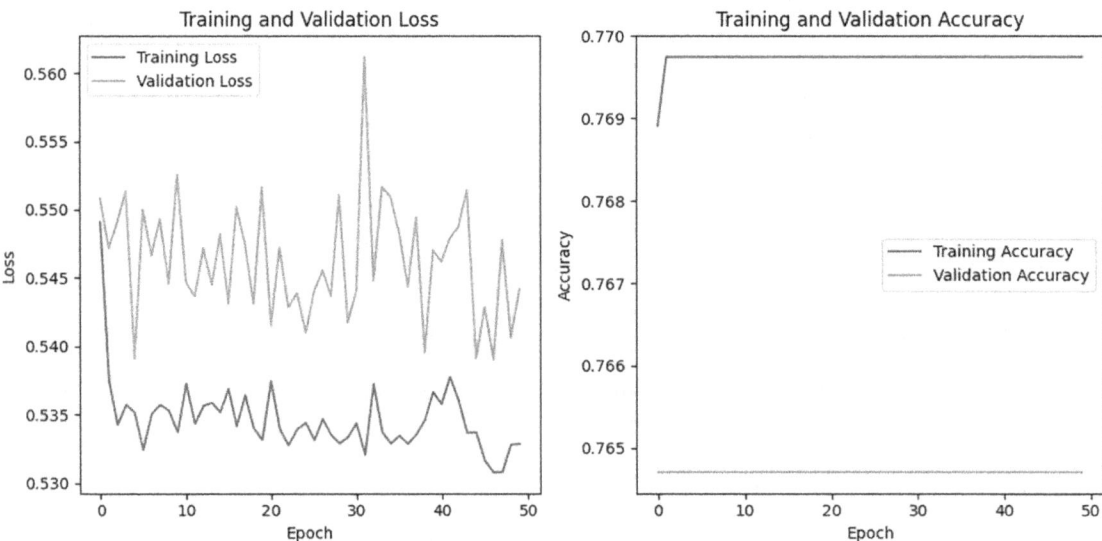

Figure 3-21. *Training and validation loss and accuracy variation with number of epochs*

CHAPTER 3   NEURAL NETWORKS

Figure 3-21 (left) shows the variation of training and validation loss with the number of epochs, and Figure 3-21 (right) shows the variation of performance with the number of epochs for model 1.

Listing 3-9. Implementing MLP with two hidden layers of 25 and 12 neurons using the Keras module to classify the Myocardial Infarction Complications dataset

Code (two hidden layers of 25 and 12 neurons):
#1. The *ucimlrep* is installed and fetched to import the *myocardial_infarction_complications* dataset.
```
!pip install ucimlrepo
from ucimlrepo import fetch_ucirepo
myocardial_infarction_complications = fetch_ucirepo(id=579)
X = myocardial_infarction_complications.data.features
y = myocardial_infarction_complications.data.targets
y = y['ZSN']
```
#2. The NaNs are calculated for each feature and dropped those having greater than threshold.
```
nan_count_per_column = X.isnull().sum()
print(nan_count_per_column)
threshold = len(X)*0.3
df = X.dropna(axis=1, thresh=threshold)
print(df)
```
#3. From *sklearn.impute* module the KNN imputer is imported to impute the remaining NaN values in the dataset.
```
import pandas as pd
from sklearn.impute import KNNImputer
imputer = KNNImputer()
df_imputed = pd.DataFrame(imputer.fit_transform(df), columns=df.columns)
print(df_imputed)
X = df_imputed
print(X.shape, y.shape)
```

#4. From *sklearn.model_selection* module the *train_test_split* function is imported to split the data into train and test.
```
X_train, X_test, y_train, y_test = train_test_split(X, y, test_size = 0.3)
print(X_train.shape, X_test.shape, y_train.shape, y_test.shape)
```
#5. The model has an input layer, two dense layers of 25 and 12 neurons (hidden layer) and a dense layer of 1 neuron (for output) with sigmoid activation is created. The model is complied with an 'sgd' optimizer, binary cross entropy loss (binary classification), and accuracy metric. The model is trained over 50 epochs with the training set.
```
model_3 = Sequential()
model_3.add(Dense(units=25, input_dim= 109, activation='sigmoid'))
model_3.add(Dense(units=12, activation='sigmoid'))
model_3.add(Dense(units=1, activation='sigmoid'))
model_3.compile(optimizer='sgd', loss='binary_crossentropy',
metrics=['accuracy'])
history = model_3.fit(X_train, y_train, epochs=50, batch_size=32,
validation_data=(X_test, y_test))
loss, accuracy = model_3.evaluate(X_train, y_train)
print(f"Loss: {loss}, Accuracy: {accuracy}")
```
#6. Note that after compiling the model the output was saved in a variable called history. This is a dictionary from which training and validation accuracy and loss are plotted.
```
import matplotlib.pyplot as plt
train_loss = history.history['loss']
val_loss = history.history['val_loss']
train_acc = history.history['accuracy']
val_acc = history.history['val_accuracy']
plt.figure(figsize=(10, 5))
plt.subplot(1, 2, 1)
plt.plot(train_loss, label='Training Loss')
plt.plot(val_loss, label='Validation Loss')
plt.xlabel('Epoch')
```

```
plt.ylabel('Loss')
plt.title('Training and Validation Loss')
plt.legend()
plt.subplot(1, 2, 2)
plt.plot(train_acc, label='Training Accuracy')
plt.plot(val_acc, label='Validation Accuracy')
plt.xlabel('Epoch')
plt.ylabel('Accuracy')
plt.title('Training and Validation Accuracy')
plt.legend()
plt.tight_layout()
plt.show()
```

Output:

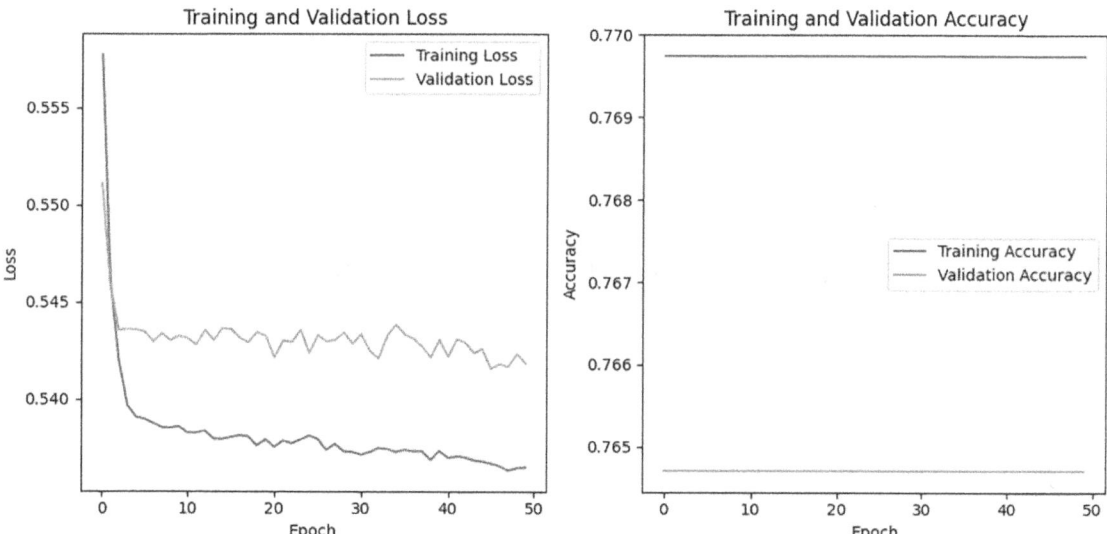

Figure 3-22. *Training and validation loss and accuracy variation with number of epochs*

CHAPTER 3   NEURAL NETWORKS

Figure 3-22 (left) shows the variation of training and validation loss with the number of epochs, and Figure 3-22 (left) shows the variation of performance with the number of epochs for model 2.

The variation of learning rate and optimizers including "SGD," "Adam," and "RMSprop" is also analyzed in Figures 3-23, 3-24, and 3-25.

Stochastic Gradient Descent

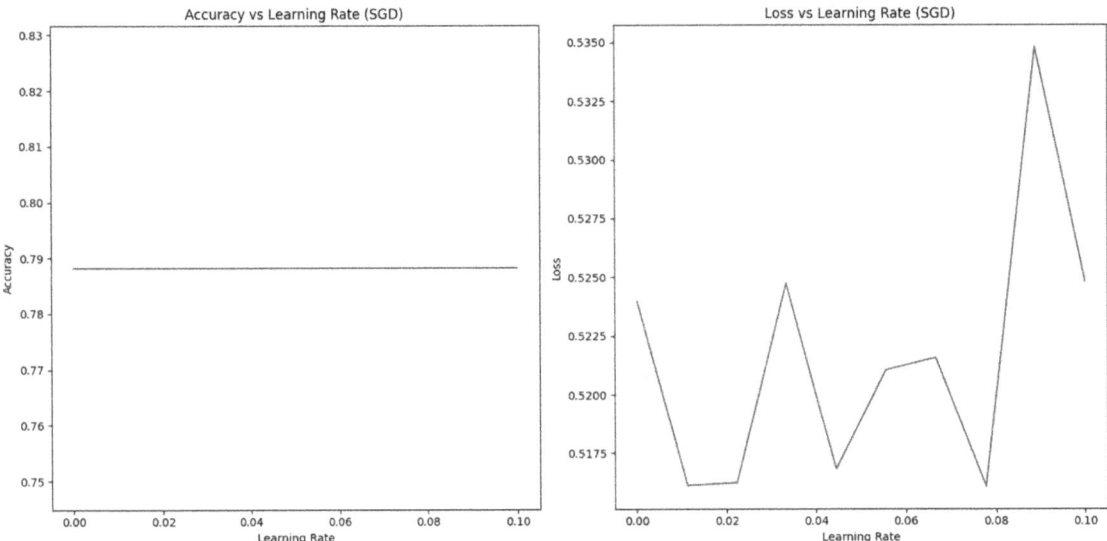

Figure 3-23. *Variation of loss and accuracy with learning rate for the stochastic gradient descent optimizer*

CHAPTER 3 NEURAL NETWORKS

RMSprop

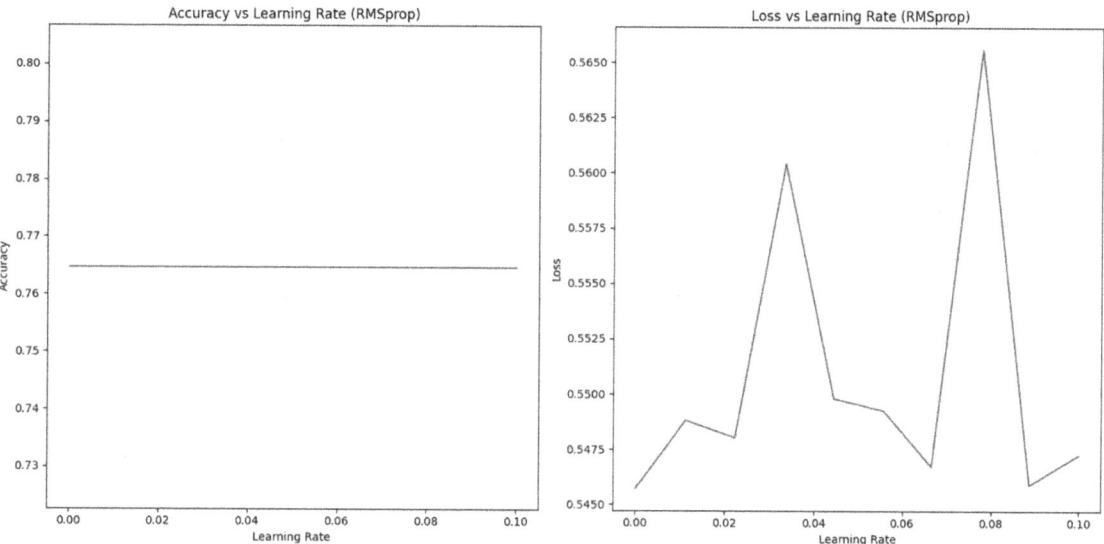

Figure 3-24. *Variation of loss and accuracy with learning rate for the RMSprop optimizer*

Adam

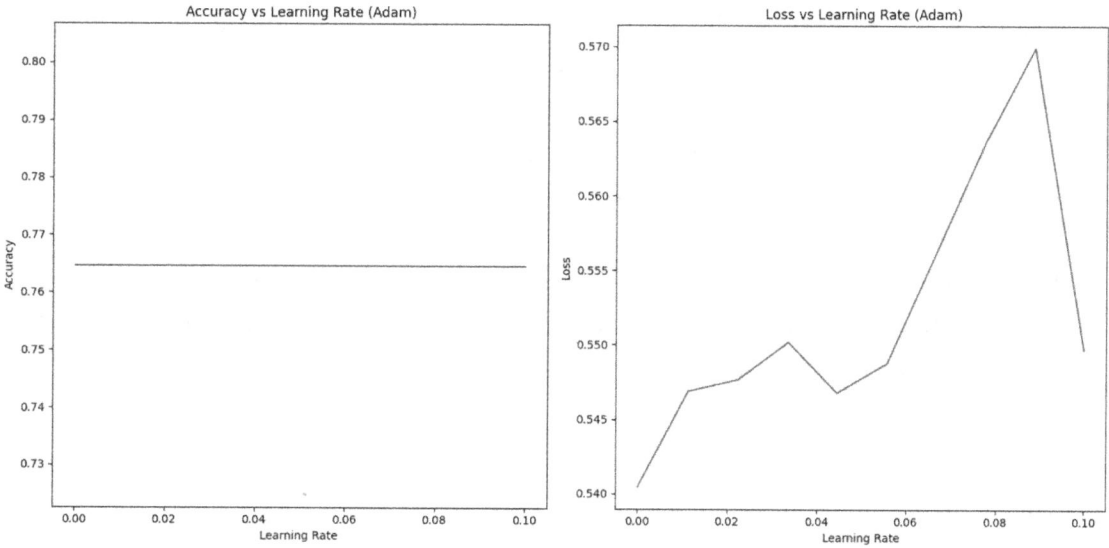

Figure 3-25. *Variation of loss and accuracy with learning rate for the Adam optimizer*

103

The results are summarized in Table 3-5.

Table 3-5. *Results of the Above Two Models on the Myocardial Infarction Complications Dataset*

MLP No.	Dataset	Model	Accuracy	Loss
1.	Myocardial Infarction Complications	model_2 (single hidden layer with 50 neurons)	0.7697	0.5315
2.		model_3 (two hidden layers with 25 and 12 neurons)	0.7697	0.5363

Note that the selection of the number of hidden layers, and the number of neurons in each layer, is a precarious task. This discussion continues in the following chapters.

Conclusion

This chapter introduced Neural Networks, which are the basis of Deep Learning models. The chapter began with an informed discussion on Single-Layer Perceptron and its limitations. It then moved to a discussion on Multi-layer Perceptron and the solution of the XOR problem. The chapter also discussed the feed-forward networks and the backpropagation algorithm for Neural Networks. Furthermore, topics such as the variation of performance with the learning rate and the depth of the network have been discussed in the chapter. The chapter includes the implementation of some important models that demonstrate the effect of these hyperparameters on the performance of the model. The next two chapters continue the discussion and introduce the reader to two important concepts, namely, bias and variance. The reader is requested to attempt the exercises to get hold of the concepts learned in the chapter.

Exercises
Multiple-Choice Questions

1. Which of the following logic gates cannot be implemented using a Single-Layer Perceptron?

    a. NAND

    b. NOR

    c. XOR

    d. All of the above

2. Which of the following can be classified using a Single-Layer Perceptron?

    a. Linearly separable data

    b. Nonlinearly separable data

    c. Both of the above

    d. None of the above

3. What is the purpose of a nonlinear activation function in a Single-Layer Perceptron?

    a. To incorporate nonlinearity to the weighted sum of input features.

    b. At times, the activation function converts the values of the input into a certain range, for example, 0 and 1.

    c. The nonlinear activation function makes the classification complex and inefficient.

    d. None of the above.

CHAPTER 3   NEURAL NETWORKS

4. Ideally what should be the primary purpose of hyperparameter tuning?

   a. To achieve better training accuracy
   b. To achieve better test accuracy
   c. To reduce the training loss
   d. To reduce the test loss

5. The sigmoid activation function is represented as $f(x) = \frac{1}{1+e^{-x}}$. What is the derivative of f in terms of f?

   a. $f(1-f)$
   b. $f(1+f)$
   c. $f(f)$
   d. None of the above

6. The sigmoid function may be represented as $f(x) = \frac{1}{1+e^{-s}}$. If the value of s is very large, the function behaves as

   a. Step function
   b. Tanh
   c. ReLU
   d. None of the above

7. In the above question, if the value of s is very small, the function behaves as

   a. Step function
   b. Tanh
   c. ReLU
   d. None of the above

8. If $f(x) = \frac{1}{1+e^{-x}}$ what is the relationship between $f(x)$ and $f(-x)$?

   a. $f(x) = 1 - f(-x)$
   b. $f(x) = 1 + f(-x)$

CHAPTER 3 NEURAL NETWORKS

c. $f(-x) = 1 - f(x)$

d. $f(-x) = 1 + f(x)$

9. In a Multi-layer Perceptron, the output of various hidden layers represents

   a. Hierarchical feature representation
   b. Outputs with different accuracy
   c. Values of the weighted inputs of each layer
   d. None of the above

10. What is the minimum number of hidden layers in a Multi-layer Perceptron needed to model any input function?

    a. 0
    b. 1
    c. 2
    d. None of the above

11. If the value of the learning rate is very small, then

    a. It takes more time to find the optimal values of the parameters of the model.
    b. It takes less time to find the optimal values of the parameters of the model.
    c. Time does not depend on learning rate.
    d. None of the above.

12. If the value of the learning rate is very large, then

    a. We may skip the optimal solution.
    b. It takes less time to find the optimal values of the parameters of the model.
    c. Time does not depend on learning rate.
    d. None of the above.

CHAPTER 3 NEURAL NETWORKS

Theory

a. Implement the following using a Single-Layer Perceptron:
   $y = \overline{A+B}$ (NOR gate) where y is the output and A and B are the inputs. You are expected to find the values of weights and the threshold for a Single-Layer Perceptron.

b. Implement the following using a Multi-layer Perceptron:
   $y = AB + \overline{AB}$ (XNOR gate) where y is the output and A and B are the inputs. You are expected to find the values of weights and the threshold for a Multi-layer Perceptron.

c. The tanh activation can be expressed as $f(x) = \dfrac{e^x - e^{-x}}{e^x + e^{-x}}$. Express the derivative of tanh with respect to tanh.

d. If $f(x) = \dfrac{e^x - e^{-x}}{e^x + e^{-x}}$, find the relationship between $f(x)$ and $f(-x)$.

e. In a Multi-layer Perceptron prove that as the number of layers increases, the use of sigmoid and tanh activation will hamper the learning of weights of earlier layers.

f. Explain the backpropagation algorithm. Derive the formula for backpropagation for a Multi-layer Perceptron having two hidden layers if the

   a. Activation function is sigmoid.

   b. Activation function is tanh.

g. Compare the features of various activation functions and explain why ReLU is considered better as compared with the rest.

h. Prove that a Single-Layer Perceptron cannot classify nonlinearly separable data.

Numerical

a. Consider two networks, one having an input layer of 128 neurons and a single hidden layer of 64 neurons and the other having 128 neurons in the input layer and two hidden layers of 32 and 16 neurons. Which do you think is better and why? Explain your answer in terms of the number of parameters and learning.

b. Consider a network having four neurons in the input layer and three neurons in the hidden layer and a single neuron in the output layer. The initial inputs, weights, and bias associated with them and the actual output are given as follows:

The given inputs $x_1 = 0.5$, $x_2 = 0.1$, $x_3 = 0.4$, $x_4 = 0.7$ and the initial random weights $w_{11} = 0.2$, $w_{12} = -0.1$, $w_{13} = 0.4$, $w_{21} = 0.5$, $w_{22} = 0.3$, $w_{23} = 0.1$, $w_{31} = -0.4$, $w_{32} = 0.2$, $w_{33} = 0.5$, $w_{41} = 0.3$, $w_{42} = -0.2$, $w_{43} = 0.2$ of input to the hidden layer and $w_{11} = 0.3$, $w_{21} = 0.2$, $w_{31} = 0.6$ of hidden to the output layer. The actual value of output is 0.6. The learning rate is 0.1.

What will be the updated weights and bias for the hidden and the output layer after the first and the second iteration?

References

[1] Finger, S. (2001). Origins of Neuroscience: A History of Explorations Into Brain Function.

[2] Rosenblatt, Frank (1957). "The Perceptron—a perceiving and recognizing automaton" (PDF). Report 85-460-1. Cornell Aeronautical Laboratory.

[3] McCulloch, W. S., & Pitts, W. (1943). A logical calculus of the ideas immanent in nervous activity. The Bulletin of Mathematical Biophysics, 5(4), 115–133. https://doi.org/10.1007/bf02478259

CHAPTER 4

Training Deep Networks

Introduction

Now that you have studied various Neural Network architectures, the gradient descent algorithm, and the backpropagation algorithm, let us explore some more optimization techniques and analyze their effect on the smoothness of the loss curve. You will also explore the effect of these techniques on the performance of the model.

In this chapter, you will study the ways of splitting a dataset and selecting an appropriate number of samples for training the network, in each iteration. You will also understand the problems in gradient descent and explore the techniques to deal with these problems. You will explore optimizers, namely, RMSprop, Momentum, and the Adam optimizer. Additionally, you will carry out an empirical analysis to study the effect of the above techniques on the performance of a network.

Train–Test Split

The objective of a ML classification model is to learn the patterns from the training data and use these patterns to classify the unseen data. The data with which the model is trained is called the training data. This data helps the model learn the parameters. The model so formed is then used to classify the data that has not been seen so far (yet unseen data). This data is called the test data. The division of data into train and test can be done in many ways. To begin with, we can simply take 70% of the data for training and the rest for testing. This number may vary.

Train–Validation–Test Split

The second method is to divide the data into three parts: a bigger part and two smaller parts. The bigger part (training set) is used to train the model and learn the parameters of the model, whereas one of the smaller parts is used to set the hyperparameters. This is called the validation set. The third part is used to test the model. For example, if you are given a sufficient amount of data, you can take 70% of the data for training the model and find the performance with the validation set. If the performance is not good, you retrain the model by changing the hyperparameters such as learning rate, number of layers, number of units in each layer, etc. When all the hyperparameters are chosen so as to optimize the performance with the validation set, then you take the test set to test the model.

K-Fold Split

In the third method, the given data is divided into "K" parts. One of the parts (say part 1) is used as a test set, whereas the other "K - 1" parts are clubbed together and used as the training set. This process is repeated "K" times by taking all the "K" parts (one at a time) as the test set. Therefore, "K" such models are developed, and the average performance of the model is reported. Figure 4-1 shows the K-fold split.

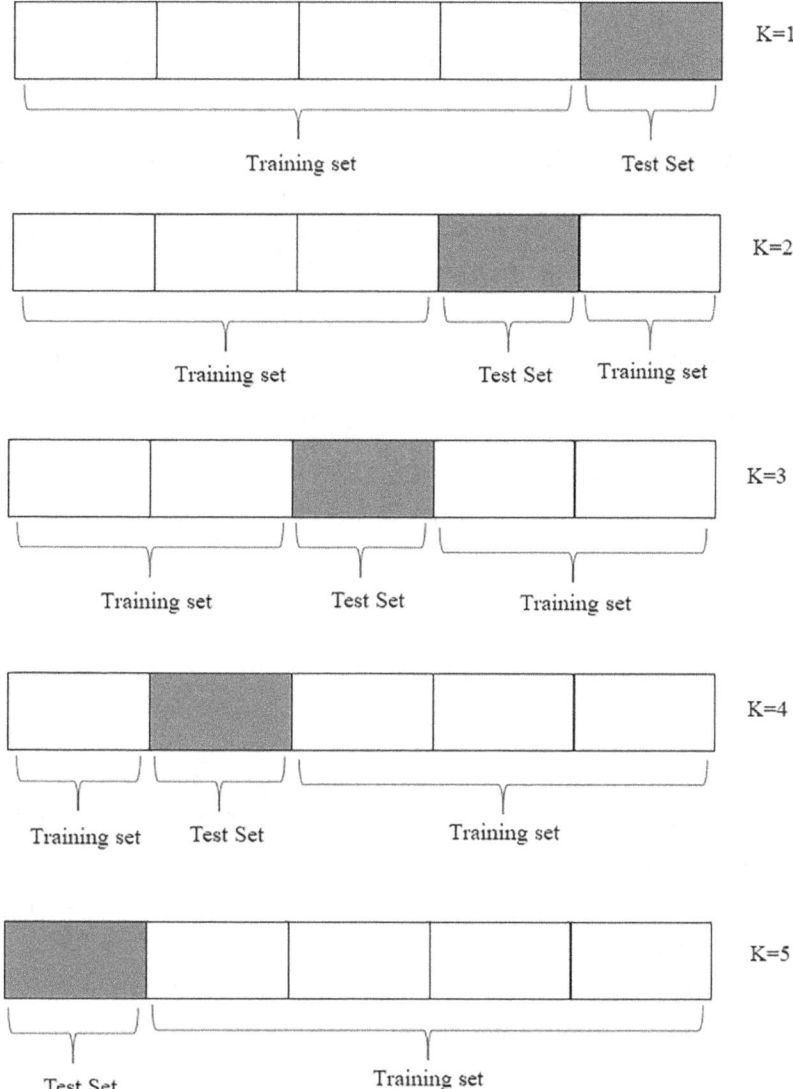

Figure 4-1. *K-fold splitting technique*

Having seen the splitting of data, let us now have a look at how many samples we should take before updating the parameters of the model.

Batch, Stochastic, and Mini-batch Gradient Descent

As stated earlier, we aim to learn the parameters of the model with the help of a training set. For this, we can either take all the samples together in a single iteration and update the weights or take one sample at a time (before updating the weights). There is also a middle path, which is to take a few samples at a time, update the weights, and then proceed further.

Batch Gradient Descent

In batch gradient descent (BGD) we process all the examples at the same time. However, if the number of examples is huge, then the training is **computationally expensive**, and the whole data set might **not fit** into the **main memory**. Therefore, we prefer stochastic or mini-batch gradient descent. The formal algorithm of batch gradient descent is as follows:

> Initialize learning rate η and parameters W.
>
> Repeat till the termination condition is met:
>
>> Find the gradient (g) over all the training examples.
>>
>> Update $W \dashrightarrow W - \eta \times g$.
>
> end while

Stochastic Gradient Descent

In stochastic gradient descent, we take one training example at a time and update the weights. This is another extreme in which we will have to wait for a long time until the whole training set is seen by the model. However, **updating** the parameters is **fast**. When the number of training examples is very large, then there can be additional overhead for the model. In this case, we **generally reach the global minima**, whereas in the case of batch gradient descent, we **might miss the global minima**.

Mini-batch Gradient Descent

In mini-batch gradient descent (mini-batch GD), we form small batches and update the parameters of the model with each batch. It is generally faster and gracefully handles the problems of batch and stochastic gradient descent. For example, if we have 1,048,576 samples in the training set and we take 1024 examples at a time, then there will be 1024 mini-batches. That is, the parameters will be updated 1024 times in iterating over the whole dataset. In this case, the loss function might not be smooth because of the fact some of the batches might be easily trainable while others may not be. Here the selection of the number of samples in a mini-batch is a hyperparameter. It should not be very

CHAPTER 4 TRAINING DEEP NETWORKS

small or very large. Generally, mini-batch gradient descent is in between batch and stochastic gradient descent both in terms of accuracy and time.

The following experiment evaluates the performance of different activation functions (sigmoid, ReLU, tanh, and a custom tanh) using three gradient descent methods (batch, mini-batch, and stochastic) with a Neural Network explained in the previous chapter on the MNIST dataset. The MNIST dataset consists of 60,000 training images and 10,000 test images of handwritten digits (0–9). The different models were created using the above stated activation functions. Each model was trained using the SGD. The training and validation accuracy and loss were plotted for each model over ten epochs as shown in Figures 4-2 and 4-3. It may be noted that the batch gradient descent had the shortest training time due to fewer updates, while the stochastic gradient descent had the longest training time due to more frequent updates. The mini-batch gradient descent provided a good balance between training time and performance.

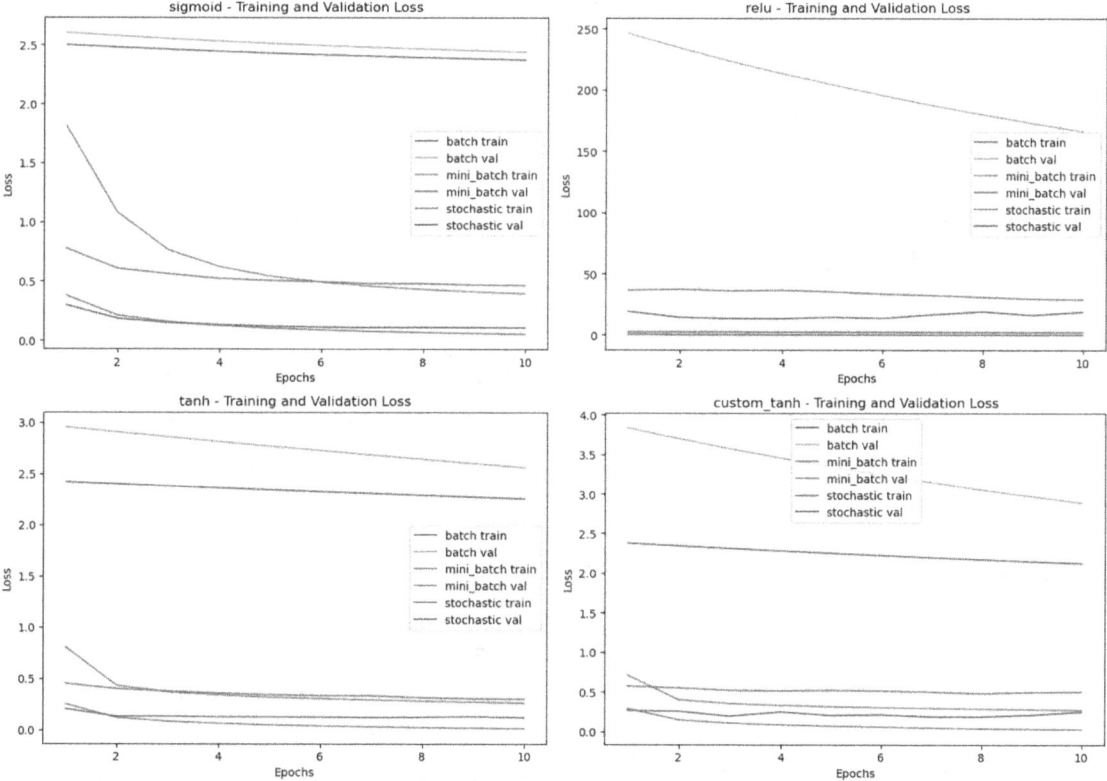

Figure 4-2. *Loss variation for different activation functions with three gradient descent methods: batch, mini-batch, and stochastic*

CHAPTER 4 TRAINING DEEP NETWORKS

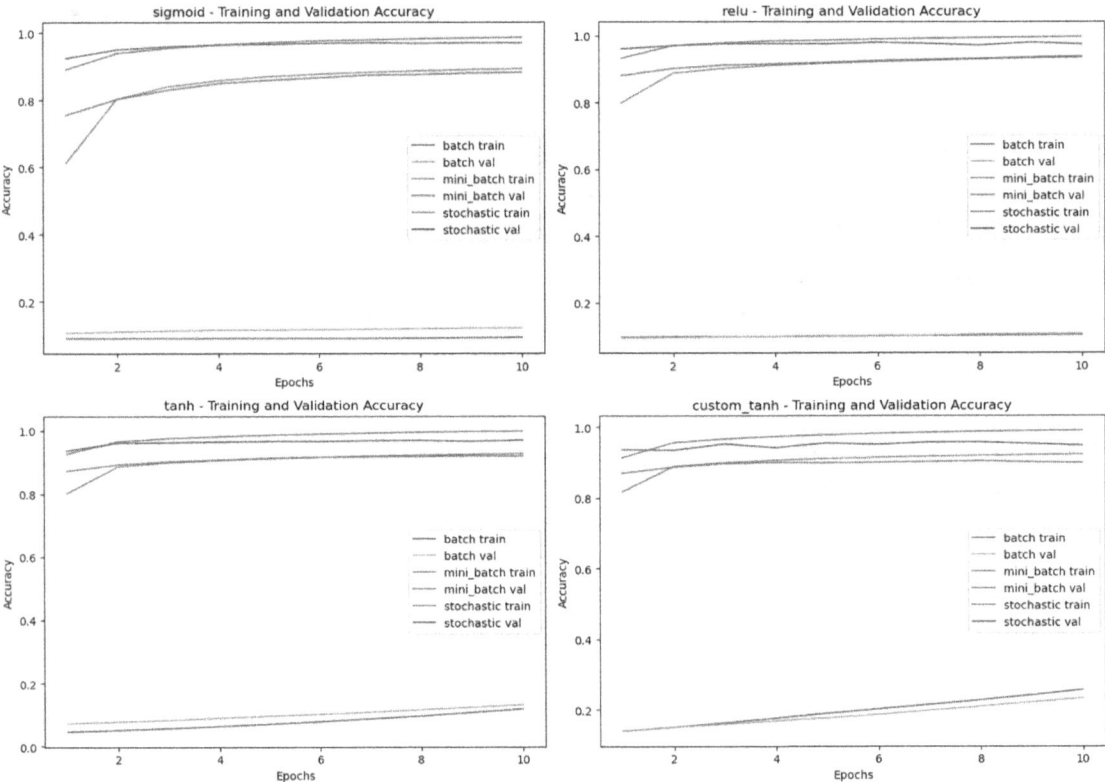

Figure 4-3. *Performance variation for different activation functions with three gradient descent methods*

Now that we have seen the division of data into train and test sets, and studied how many samples should be taken before updating the weights of the model, let us now have a look at some important optimization methods. We begin with RMSprop. Also, we will study one of the most important optimization methods: Adam optimizer.

RMSprop

In the case of gradient descent, the initial weights and bias are updated in each iteration with the aim of minimizing the loss. However, the variation of loss with iterations may not be smooth. If we have a single weight and bias, then with each iteration the bias is updated, and this variation is shown in one of the axes (say Y), whereas the variation of weight is reflected in another axis (say X). The overall variation can be seen in Figure 4-5.

CHAPTER 4   TRAINING DEEP NETWORKS

Now we aim to slow the learning in the Y-axis, whereas keep the learning as good as earlier in the X-axis; we can make slight changes in the formulas that update the weight and bias. Kindly note that the notations used in the following algorithms are the same as the course slides (optimization algorithms) of DeepLearning.AI [1].

After each update, divide d_w and d_b, respectively, by $\sqrt{S_{dw}}$ and $\sqrt{S_{db}}$. Here S_{db} is large in comparison with S_{dw}, and hence the change in weight in the Y-axis is small as compared with earlier. Here S_{dw} is the weighted average of the earlier S_{dw} and dw^2. Likewise, S_{db} can also be considered as the weighted average of S_{db} and db^2. Here we have a parameter β that may be considered a hyperparameter. That is, first of all, we initialize the following parameters:

- Learning rate (α).
- Decay rate (β).
- Small constant (ϵ).
- Initialize S_{dw} and S_{db} to zero.

This is followed by the application of the following algorithm to update the weights in each iteration.

In each iteration

- Calculate d_w and d_b.
- Update the running average of the squared gradients:
  - $S_{dw} = \beta\, S_{dw} + (1 - \beta)dw^2$
  - $S_{db} = \beta\, S_{db} + (1 - \beta)db^2$
- Update the parameters:
  - $w = w - \alpha \dfrac{d_w}{\sqrt{S_{dw}} + \epsilon}$
  - $b = b - \alpha \dfrac{d_b}{\sqrt{S_{db}} + \epsilon}$

RMSprop works better as compared with Momentum in the case of non-convex settings. The algorithm was suggested by G. Hinton in one of the Coursera courses. The algorithm that follows engulfs the good parts of both Momentum and RMSprop.

CHAPTER 4   TRAINING DEEP NETWORKS

Adam Optimizer

The Adam optimizer combines the concepts of Momentum and RMSprop. It calculates v_{dw}, v_{db}, S_{dw}, and S_{db} in the same way, as explained above. Initially, the values of these four can be taken as zero, and in each iteration v_{dw} and v_{db} are calculated using the following equations:

$$v_{dw} = \beta_1 v_{dw} + (1-\beta_1)dw$$

$$v_{db} = \beta_1 v_{db} + (1-\beta_1)db$$

Likewise, S_{dw} and S_{db} can be calculated as follows:

$$S_{dw} = \beta_2 S_{dw} + (1-\beta_2)dw^2$$

$$S_{db} = \beta_2 S_{db} + (1-\beta_2)db^2$$

Now we fix the bias using the following equations:

$$v_{dw}^{corrected} = \frac{v_{dw}}{1-\beta_1^t}$$

$$v_{db}^{corrected} = \frac{v_{db}}{1-\beta_1^t}$$

$$S_{dw}^{corrected} = \frac{S_{dw}}{1-\beta_1^t}$$

$$S_{db}^{corrected} = \frac{S_{db}}{1-\beta_1^t}$$

Now the weights will be updated using the above calculated values:

$$w = w - \alpha \frac{v_{dw}^{corrected}}{\sqrt{S_{dw}^{corrected}} + \epsilon}$$

$$b = b - \alpha \frac{v_{db}^{corrected}}{\sqrt{S_{db}^{corrected}} + \epsilon}$$

Here, we have three hyperparameters α, β_1, and β_2. The learning rate can be estimated using various methods like grid search and heuristic algorithms. β_1 is the parameter for Momentum and β_2 is the parameter of RMSprop. As per Krohn [2], generally, the values of β_1 and β_2 are taken as 0.9 and 0.99. The formal algorithm for the Adam optimizer is as follows:

Algorithm: Adam Optimizer

Initialize parameters:

- Learning rate (α).
- Decay rates (β_1 and β_2).
- Small constant (ϵ).
- Initialize v_{dw} and S_{dw} to zero.
- Initialize v_{db} and S_{db} to zero.

In each iteration

- Calculate the gradients d_w and d_b.
- Update biased first moment estimates:
  - $v_{dw} = \beta_1 v_{dw} + (1 - \beta_1)dw$
  - $v_{db} = \beta_1 v_{db} + (1 - \beta_1)db$
- Update biased second moment estimates:
  - $S_{dw} = \beta_2 S_{dw} + (1 - \beta_2)d_w^2$
  - $S_{db} = \beta_2 S_{db} + (1 - \beta_2)d_b^2$
- Compute bias-corrected first moment estimates:
  - $v_{dw}^{corrected} = \dfrac{v_{dw}}{1 - \beta_1^t}$
  - $v_{db}^{corrected} = \dfrac{v_{db}}{1 - \beta_1^t}$
- Compute bias-corrected second moment estimates:
  - $S_{dw}^{corrected} = \dfrac{S_{dw}}{1 - \beta_1^t}$
  - $S_{db}^{corrected} = \dfrac{S_{db}}{1 - \beta_1^t}$

CHAPTER 4   TRAINING DEEP NETWORKS

- Update the parameters:

  - $w = w - \propto \dfrac{v_{dw}^{corrected}}{\sqrt{S_{dw}^{corrected}} + \epsilon}$

  - $b = b - \propto \dfrac{v_{db}^{corrected}}{\sqrt{S_{db}^{corrected}} + \epsilon}$

To understand the variations of the loss function with different optimizers such as gradient descent, RMSprop, and Adam, let us take a very simple example. The popular IRIS dataset has four features and three classes, out of which the first two classes are taken. Initially, the weights are set to small random numbers, and they are updated in each iteration using three different techniques stated above. Let us explore the variation of loss in each epoch. We vertically concatenate the weights in each epoch and then apply PCA to take the first component having maximum variance. Refer to Listing 4-1. The variation of weight (X-axis) and bias (Y-axis) with the number of epochs is shown in Figures 4-4, 4-5, and 4-6.

Listing 4-1. Variation of weights and bias for different optimization techniques

Code:
#1. Import the requisite packages
```
import numpy as np
import matplotlib.pyplot as plt
from sklearn.datasets import load_iris
from sklearn.model_selection import train_test_split
from mpl_toolkits.mplot3d import Axes3D
```
#2. Load the IRIS dataset and take the first 100 samples
```
iris = load_iris()
X = iris.data[:100]  # Select only the first two classes for binary classification
y = iris.target[:100].reshape(-1, 1)  # Reshape to column vector
```
#3. Split dataset
```
X_train, X_test, y_train, y_test = train_test_split(X, y, test_size=0.3, random_state=42)
```
#4. Set the hyperparameters for Adam Optimizer
```
np.random.seed(42)
w = np.random.randn(X_train.shape[1], 1)
```

```python
b = np.random.randn(1)
α = 0.01
initial_w = w.copy()
initial_b = b.copy()
# Hyperparameters for Adam
β1 = 0.9
β2 = 0.999
ε = 1e-8
```
#5. Initialize the variables of the Adam optimizer
```python
v_dw = np.zeros_like(w)
S_dw = np.zeros_like(w)
v_db = 0
S_db = 0
```
#6. Implement the Sigmoid function
```python
def sigmoid(z):
    return 1 / (1 + np.exp(-z))
```
#7. Compute gradients
```python
def compute_gradients(X, y, w, b):
    m = X.shape[0]
    y_pred = sigmoid(np.dot(X, w) + b)
    dw = (1/m) * np.dot(X.T, (y_pred - y))
    db = (1/m) * np.sum(y_pred - y)
    return dw, db
```
#8. Update parameters using Adam
```python
def update_adam(w, b, dw, db, t, α, v_dw, S_dw, v_db, S_db, β1=0.9, β2=0.999, ε=1e-8):
```
    #9. Update biased first moment estimates
```python
    v_dw = β1 * v_dw + (1 - β1) * dw
    v_db = β1 * v_db + (1 - β1) * db
```
    #10. Update biased second moment estimates
```python
    S_dw = β2 * S_dw + (1 - β2) * (dw ** 2)
    S_db = β2 * S_db + (1 - β2) * (db ** 2)
```
    #11. Compute bias-corrected first moment estimates
```python
    v_dw_corrected = v_dw / (1 - β1 ** t)
    v_db_corrected = v_db / (1 - β1 ** t)
```

#12. Compute bias-corrected second moment estimates
```
S_dw_corrected = S_dw / (1 - β2 ** t)
S_db_corrected = S_db / (1 - β2 ** t)
```
#13. Update parameters
```
w -= α * v_dw_corrected / (np.sqrt(S_dw_corrected) + ϵ)
b -= α * v_db_corrected / (np.sqrt(S_db_corrected) + ϵ)
return w, b, v_dw, S_dw, v_db, S_db
```
#14. Carry out Training
```
num_epochs = 100
weight_updates_adam = []
bias_updates_adam = []
w_adam = initial_w.copy()
b_adam = initial_b.copy()
for epoch in range(num_epochs):
    t = epoch + 1
    dw, db = compute_gradients(X_train, y_train, w_adam, b_adam)
    w_adam, b_adam, v_dw, S_dw, v_db, S_db = update_adam(w_adam, b_adam, dw, db, t, α, v_dw, S_dw, v_db, S_db, β1, β2, ϵ)
    weight_updates_adam.append(w_adam.copy())
    bias_updates_adam.append(b_adam.copy())
```
#15. Plot the variation of w and b with the number of epochs
```
def plot_3d(weight_updates, bias_updates, title):
    fig = plt.figure()
    ax = fig.add_subplot(111, projection='3d')
    epochs = range(1, num_epochs + 1)
    weight_updates_flat = np.array(weight_updates).reshape(num_epochs, -1)
    bias_updates_flat = np.array(bias_updates).reshape(num_epochs, -1)
    ax.plot(epochs, weight_updates_flat[:, 0], bias_updates_flat[:, 0], label='Weight and Bias updates')
    ax.set_xlabel('Epoch')
    ax.set_ylabel('Weight Component')
    ax.set_zlabel('Bias')
    ax.set_title(f'{title} Weight and Bias Updates')
    ax.legend()
    plt.show()
```

CHAPTER 4 TRAINING DEEP NETWORKS

```
plot_3d(weight_updates_sgd, bias_updates_sgd, 'SGD')
plot_3d(weight_updates_rmsprop, bias_updates_rmsprop, 'RMSprop')
plot_3d(weight_updates_adam, bias_updates_adam, 'Adam')
```

Output:

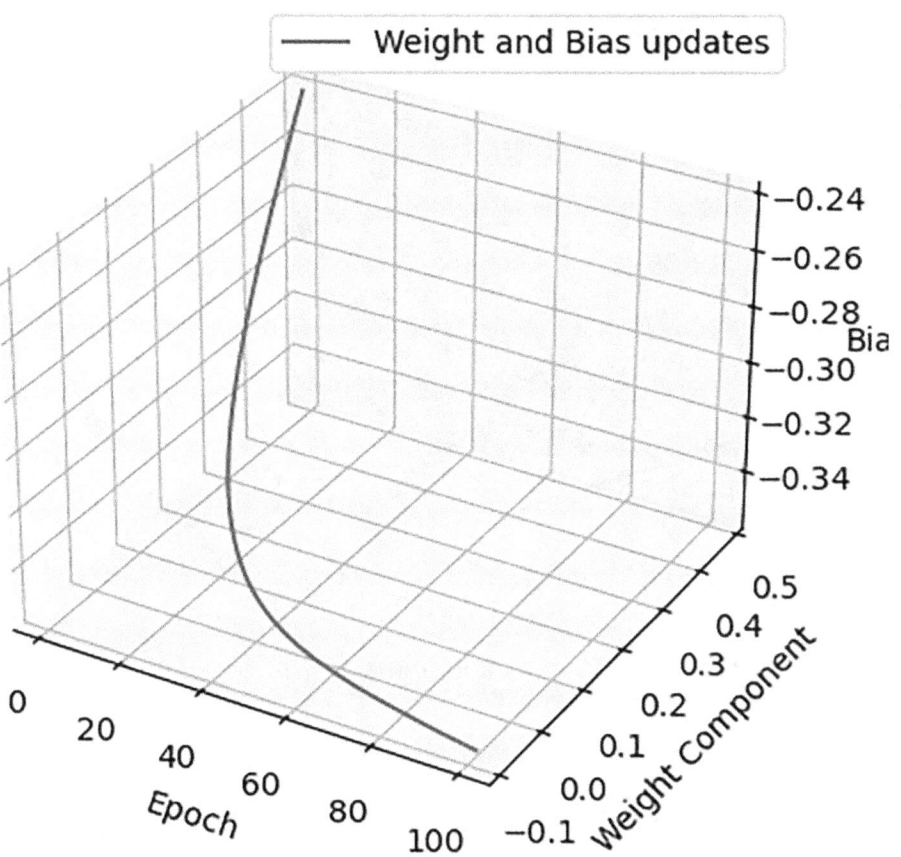

Figure 4-4. *Variation of bias and weight with number of epochs for the SGD optimizer*

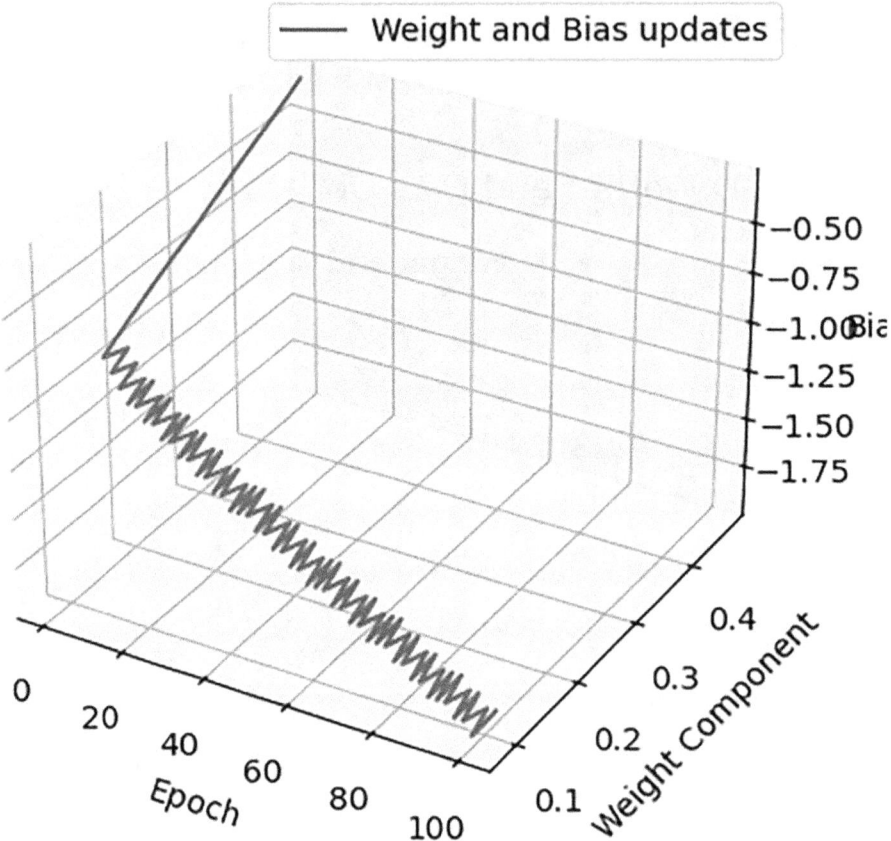

Figure 4-5. *Variation of bias and weight with the number of epochs for the RMSprop optimizer*

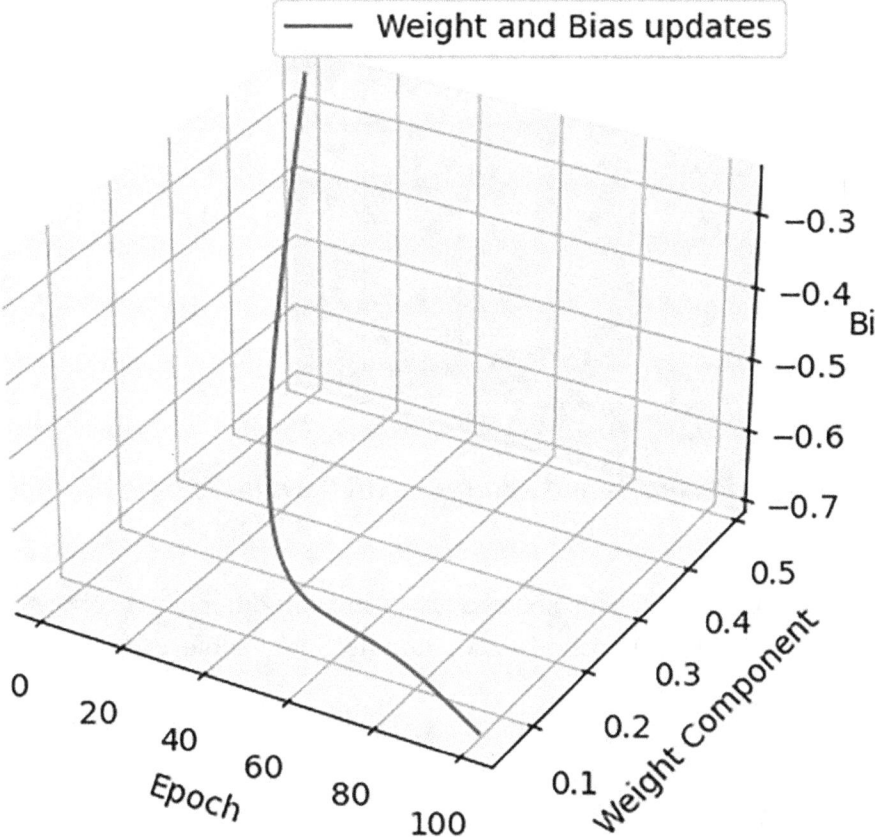

Figure 4-6. *Variation of bias and weight with number of epochs for the Adam optimizer*

The reader is requested to visit the Chapter 1. The chapter contains an implementation to classify the MNIST dataset using Neural Networks. With reference to that, the following graphs compare the loss and performance of three popular optimization algorithms, namely, SGD, RMSprop, and Adam, using a network trained on the MNIST dataset. The size of each grayscale image in the dataset was 28 × 28 pixels. A simple network was employed with an input layer of 784 units (28 × 28 pixels), followed by a hidden layer of 128 neurons with ReLU activation and an output layer of 10 neurons with softmax activation. The model was trained separately using SGD, RMSprop, and Adam optimizers for 50 epochs. The training and validation loss and accuracy are then plotted for each optimizer as shown in Figure 4-7.

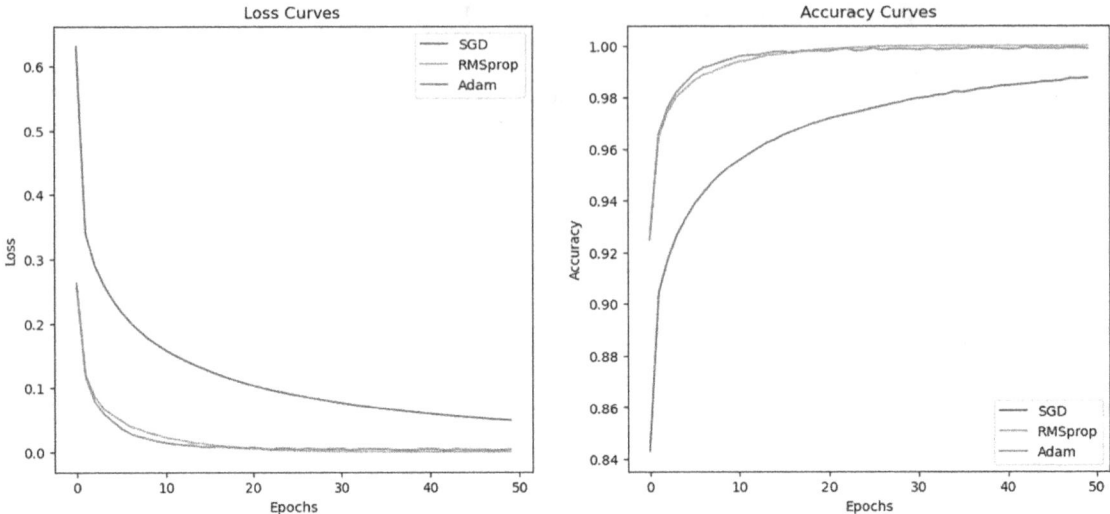

Figure 4-7. *Variation of loss and accuracy with the number of epochs for different optimizers*

Note that Adam and RMSprop showed faster and smoother convergence of the loss curve compared with SGD. Also, both Adam and RMSprop achieved higher accuracy than SGD.

Conclusion

In the last chapters, the fundamentals of Neural Networks were discussed. Since we need to create deeper models as we proceed, it is important to know the best practices of (a) dividing the data for training the model and testing and (b) finding the number of training examples that should be considered before updating the weights of the model (c) to be able to work with better optimizers, vis-à-vis stochastic gradient descent, for achieving better performance [3-5]. This chapter opens the door to the exciting world of efficient and effective Deep Neural Networks. The discussion continues in the next chapter, where we will study the concepts of bias and variance and study ways to deal with them.

Exercises

Multiple-Choice Questions

1. Which of the following techniques of updating the weights of a network may not work if the main memory is limited?

    a. Batch gradient descent (BGD).

    b. Mini-batch gradient descent (mini-batch GD).

    c. Stochastic gradient descent (SGD).

    d. All the above perform in the same manner.

2. Which of the following finds the gradient of the cost function with the parameters for the complete training set?

    a. Batch gradient descent (BGD).

    b. Mini-batch gradient descent (mini-batch GD).

    c. Stochastic gradient descent (SGD).

    d. All the above perform in the same manner.

3. Which of the following has smoother convergence on a convex landscape?

    a. Batch gradient descent (BGD).

    b. Mini-batch gradient descent (mini-batch GD).

    c. Stochastic gradient descent (SGD).

    d. All the above perform in the same manner.

4. Which of the following requires a large amount of main memory, in case of huge datasets, and otherwise may not work?

    a. Batch gradient descent (BGD).

    b. Mini-batch gradient descent (mini-batch GD).

    c. Stochastic gradient descent (SGD).

    d. All the above perform in the same manner.

CHAPTER 4 TRAINING DEEP NETWORKS

5. Which of the following can escape local minima more effectively, still better than SGD in many aspects?

    a. Batch gradient descent (BGD).

    b. Mini-batch gradient descent (mini-batch GD).

    c. Stochastic gradient descent (SGD).

    d. All the above perform in the same manner.

6. Which of the following takes a large time before the complete training data is seen by the model?

    a. Batch gradient descent (BGD).

    b. Mini-batch gradient descent (mini-batch GD).

    c. Stochastic gradient descent (SGD).

    d. All the above perform in the same manner.

7. Which of the following is the fastest of the three methods, especially for large datasets?

    a. Batch gradient descent (BGD).

    b. Mini-batch gradient descent (mini-batch GD).

    c. Stochastic gradient descent (SGD).

    d. All the above perform in the same manner.

8. Which of the following may lead to very noisy updates, making convergence slower?

    a. Batch gradient descent (BGD).

    b. Mini-batch gradient descent (mini-batch GD).

    c. Stochastic gradient descent (SGD).

    d. All the above perform in the same manner.

CHAPTER 4 TRAINING DEEP NETWORKS

9. Which of the following needs careful selection of the learning rate to avoid overshooting the minimum?

    a. Batch gradient descent (BGD).

    b. Mini-batch gradient descent (mini-batch GD).

    c. Stochastic gradient descent (SGD).

    d. All the above perform in the same manner.

10. Which of the following should be ideal batch sizes in mini-batch gradient descent?

    a. Not too large, and powers of 2

    b. Not too large, and powers of 10

    c. Large, and powers of 2

    d. Large, and powers of 10

11. Which of the following methods of dividing data into train and test sets may be preferred to handle the effect of variance in reporting the performance?

    a. Divide data into two parts: 70% for training and 30% for test.

    b. Divide data into two parts: 50% for training and 50% for test.

    c. K-fold split.

    d. None of the above.

Theory

1. Explain the problems in gradient descent and discuss how can we solve these problems.

2. Write the algorithm for updating weights using RMSprop and how can we handle the problems of Momentum.

3. Write the algorithm for updating weights using the Adam optimizer and explain how it can handle the problems of both Momentum and RMSprop.

Experiments

Take the MNIST dataset (https://keras.io/api/datasets/mnist/) and develop a Deep Neural Network having two hidden layers and ten neurons in the output layer. You may choose the number of neurons in the hidden layers by conducting various experiments. Report the performance of the model in the following cases:

1. Take the optimizer as
   a. RMSprop
   b. Adam
2. Repeat the above experiments using
   a. Stochastic gradient descent
   b. Batch gradient descent
   c. Mini-batch gradient descent
3. Vary the learning rate in all the above experiments, and find the optimal learning rate.

Plot the loss curve in each of the above cases and analyze the results.

References

[1] DeepLearning.AI. (2022, October 19). *Resources - DeepLearning.AI.* https://www.deeplearning.ai/resources/#course-slides

[2] Johnson, J. (2019). *Lecture 4: Optimization.* https://web.eecs.umich.edu/~justincj/slides/eecs498/498_FA2019_lecture04.pdf

[3] Trivedi, S., Kondor, R., & University of Chicago. (2017). Lecture 6 Optimization for Deep Neural Networks. In *CMSC 35246: Deep Learning.* https://home.ttic.edu/~shubhendu/Pages/Files/Lecture6_pauses.pdf

[4] Leal-Taixé, Prof., & Niessner, Prof. (n.d.). *Lecture 5 recap.* https://dvl.in.tum.de/slides/i2dl-ws18/6.Optimization2.pdf

[5] Sun, R., Hong, M., & Wang, J. (2019). *Lecture Notes for CIE6128: Understanding Deep Learning from a Theoretical Perspective* (By University of Illinois, University of Minnesota, & CUHK(SZ)). https://walterbabyrudin.github.io/Notes/CIE6128.pdf

CHAPTER 5

Hyperparameter Tuning

Introduction

In the previous chapters, we have discussed the architecture of Neural Networks, the gradient descent algorithm, backpropagation, and various optimization algorithms along with how to split the data for training and testing. We have also explored the effect of activation functions on the performance of the model.

The performance of the model, measured during training, does not tell us much about how it is going to perform during testing. For this, we generally find the performance of the trained model on the validation set. If the performance on the validation set is not up to the mark or less than that in the training, we change the values of the hyperparameters to handle this situation. This is called hyperparameter tuning.

Effectively, we try to reduce the variance of the model by setting the hyperparameters. In this chapter, we begin with revisiting the concepts of bias and variance and then move to hyperparameters of various architectures in Deep Learning. We will see the effect of these hyperparameters on a Deep Neural Network (DNN) in the next sections. The variation of performance with the values of hyperparameters in the case of a Convolutional Neural Network (CNN) and sequence models are discussed in the following chapters. The chapter has been organized as follows. Section "Bias–Variance Revisited" of this chapter revisits bias and variance. Section "Hyperparameter Tuning" discusses the hyperparameters of DNN, CNN, sequence models, and autoencoders, respectively. The next section, "Experiments: Hyperparameter Tuning," presents some of the experiments to empirically establish the above points and the last section concludes.

CHAPTER 5 HYPERPARAMETER TUNING

Bias–Variance Revisited

Assume that there are ten points lying on a sinusoidal curve as shown in Figure 5-1. However, there is no way to know the underlying curve; we can only see the points. We start fitting the following degree curves on these points:

- Degree 0
- Degree 1
- ...
- Degree 3

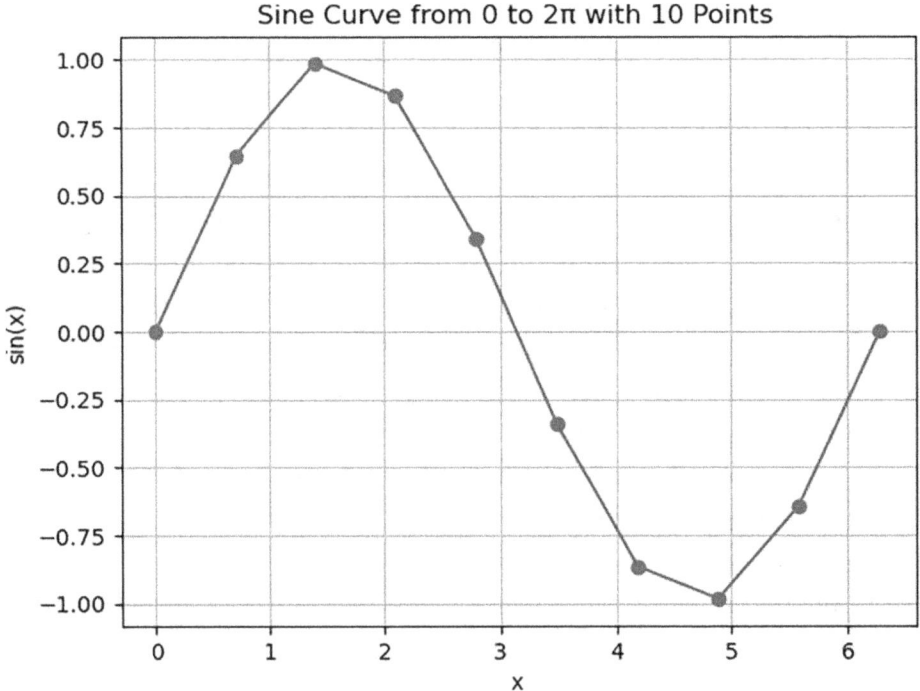

Figure 5-1. *Sinusoidal curve*

So fitting a curve having degree 1 (line) is the same as developing a linear regression model that will find out a line having the least squared distance from all the points (Figure 5-2). Likewise, nonlinear regression can create better fits on the training data. For the above points, a degree 3 curve may result in a better fit (Figure 5-3), and a degree 10 curve may result in best fit (Figure 5-4).

CHAPTER 5   HYPERPARAMETER TUNING

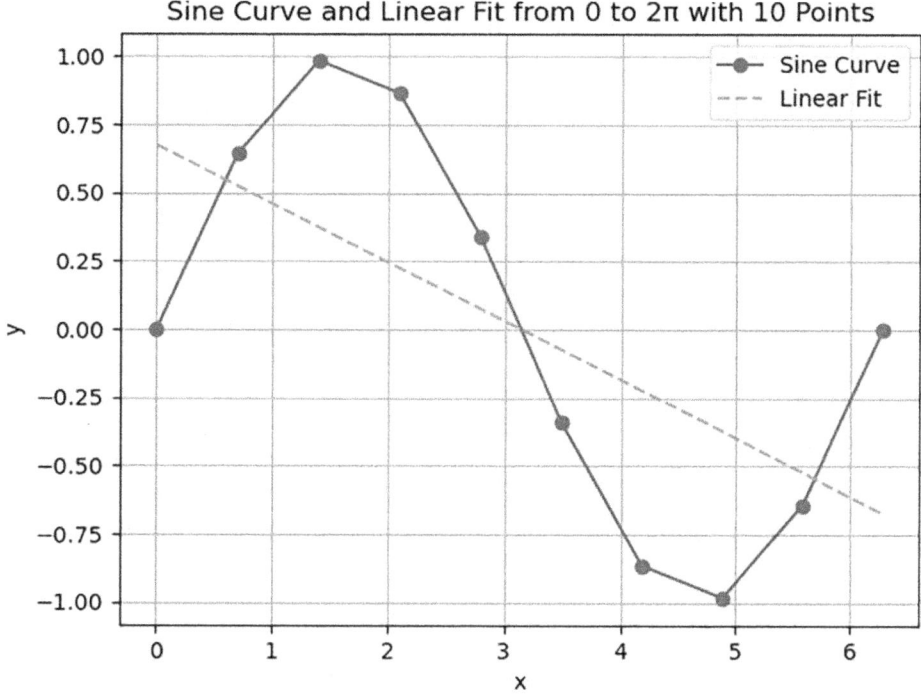

Figure 5-2. *Fitting a line to the given points*

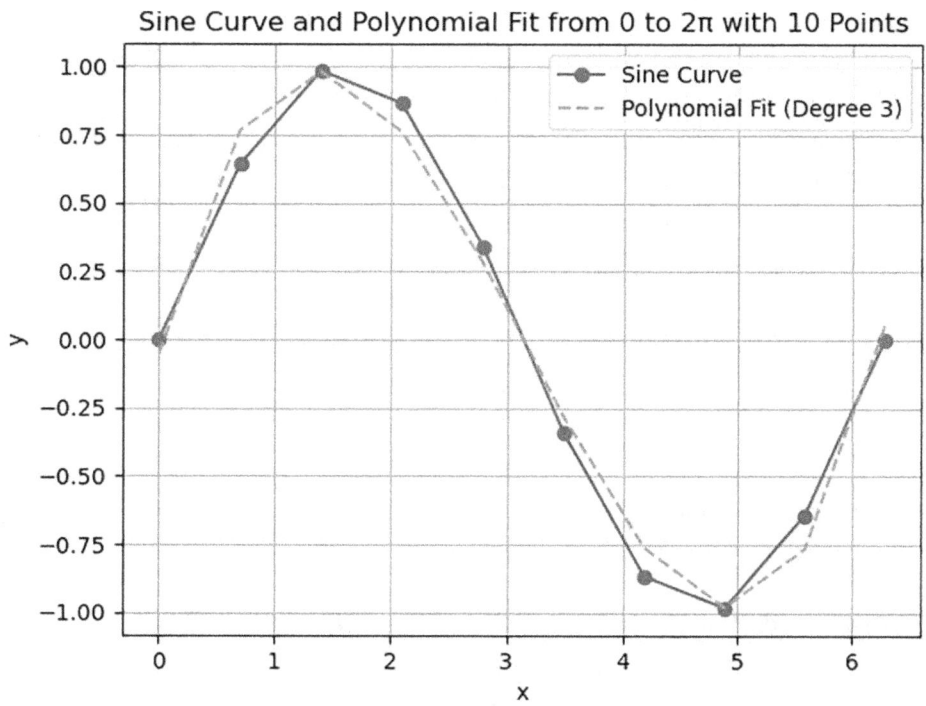

Figure 5-3. *Fitting a degree 3 curve to the given points*

135

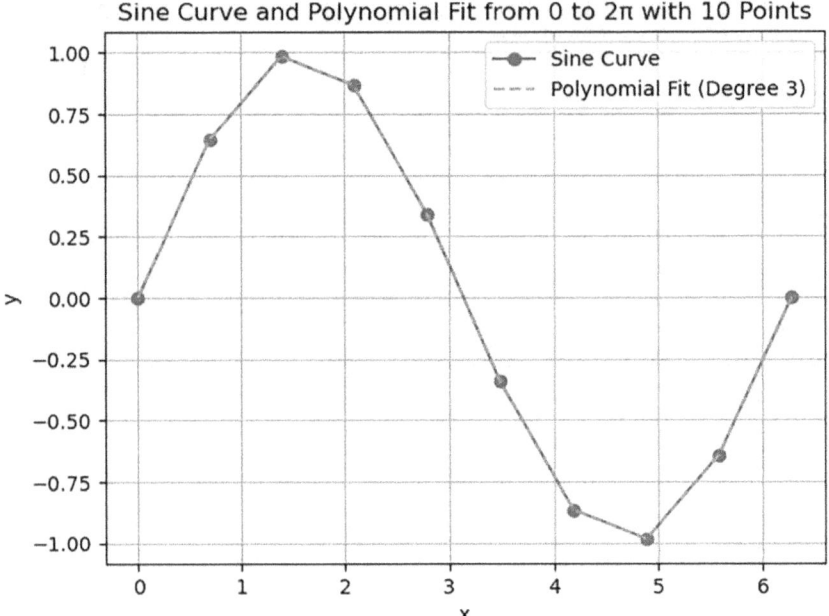

Figure 5-4. *Fitting a degree 10 curve to the given points*

Though we have been able to fit all the given points using a curve of a higher degree, the problem starts here. This is because fitting the given data (training set) is not the goal. The goal is to design a model that is able to extract the underlying structure of the given distribution to handle the unseen data points. Therefore, in the case of a curve having degree 1, both the test and the train error will be high. The model will not be able to fit either the train data or the test data. In the case of degree 2, the model may not produce a very large error with the unseen data. However, in the case of a nonlinear regression with degree 10, the training error can be very low, but the test error can be very large. So a line of best fit becomes the case of underfitting, and a curve of degree 10 will be a case of overfitting.

Tip

Overfitting: *If the training error is very low and the test error is very large, then the model is said to overfit.*

Underfitting: *If the training error is high and so is the test error like in the case of linear regression, this is called underfitting.*

CHAPTER 5   HYPERPARAMETER TUNING

In the first case (degree 1), we assumed a straight line would be able to fit the train data and predict the test data. We do not know the underlying curve, and hence we assumed that the points lie on a line (of best fit), and we would be able to find the value of y for an unseen value of x. In our example, our hypotheses were incorrect as a straight line cannot fit all points lying on a sinusoidal curve. This is called bias.

Bias   The average prediction of a good Machine Learning model should be as close to the ground truth as possible. This difference is referred to as bias.

Bias can be perceived as the ability of the underlying model to predict values. The formal definition of bias is as follows:

$$Bias = E\left[f'(x) - f(x)\right],$$

where $f(x)$ is the average predicted value of the model and $f(x)$ is the underlying function. High bias indicates the inability of the model to fit the training data. One of the reasons for this may be an oversimplified model. High bias leads to a higher error rate both with the train and the test set.

Variance   The **variance** of a model signifies its ability to adjust to a given dataset. This variability is referred to as variance.

The formal definition of variance is as follows:

$$Variance = E\left[f'(x) - f(x)\right]^2.$$

Hyperparameter Tuning

Hyperparameter tuning will partially help us deal with the problems discussed in the last section. This section presents some of the most important hyperparameters of four types of networks, namely, DNN, CNN, sequence models, and autoencoders. We begin with discussing the hyperparameters of DNN as shown in Table 5-1.

CHAPTER 5　HYPERPARAMETER TUNING

Table 5-1. Hyperparameters of DNN

Network	Image	Hyperparameters	Description
Deep Neural Network		Number of hidden layers	• The Deep Neural Network must have at least one hidden layer except for the input layer and the output. • If the number of layers is too large, the learning will be slow (vanishing gradient). • At times, depth is required to extract the hierarchy of features.
		Number of neurons in each hidden layer	If the number of hidden layers in a network is fewer and the number of neurons in that layer is more, then we generally prefer to increase the number of layers by a small amount and reduce the number of neurons in each layer.
		Learning rate	• The learning rate controls the step size of the gradient descent update. • A lower learning rate results in more time to reach optimal value. • A higher learning rate may lead to skipping the optimal value in the loss landscape.
		Batch size	The batch size indicates the number of samples processed before the model parameters are updated.
		Number of epochs	The epoch denotes the number of times the entire dataset is passed through the network

Optimizer	The selection of the algorithm used to update weights affects the recital of the model. Some of the famous optimizers are as follows: • SGD • Momentum • RMSprop • Adam
Loss function	The metric used to evaluate the performance of the model.
Activation function	"An activation, or activation function, for a neural network is defined as the mapping of the input to the output via a nonlinear transform function at each 'node,' which is simply a locus of computation within the net" [1].
Regularization	"Regularization trades a marginal decrease in training accuracy for an increase in generalizability" [2].
Dropout rate	The dropout rate denotes the fraction of the units to drop during training.

CHAPTER 5   HYPERPARAMETER TUNING

In Chapters 6 and 7 CNN is presented, which handles the task related to imaging data gracefully. The hyperparameters of this network are presented in Table 5-2.

Table 5-2. *Hyperparameters of CNN*

Network	Image	Hyperparameters	Description
Convolutional Neural Networks		Number of filters	The number of filters represents the number of convolutional filters in each layer.
		Filter size	The filter size corresponds to the dimensions of the convolutional filters, such as 3×3, 5×5, etc.
		Stride	The stride signifies the step size of the filter during convolution.
		Padding	The padding represents whether and how the input is padded, for example, valid or same.
		Pooling size	The pooling size is the dimensions of the pooling operation, for example, 2×2 and so on.
		Pooling type	The type of pooling operation, for example, max pooling, average pooling, etc.
		Dropout rate	The dropout rate denotes the fraction of the units to drop during training.

CHAPTER 5 HYPERPARAMETER TUNING

Chapter 9 and Chapter 10 of this book present sequence models. The hyperparameters of these networks are presented in Table 5-3.

Table 5-3. *Hyperparameters of Sequence Models*

Network	Image	Hyperparameters	Description
Recurrent Neural Networks (RNNs) and variants (LSTM, GRU)		Hidden units	The number of units in the RNN cell.
		Sequence length	The sequence length represents the length of the input sequences.
		Dropout rate	The dropout rate denotes the fraction of the units to drop during training.
		Number of layers	The number of stacked RNN layers.
		Learning rate	The learning rate controls the step size of the gradient descent update.
		Batch size	The batch size indicates the number of samples processed before the parameters of the model are updated.

Chapter 11 of this book discusses autoencoders. The hyperparameters of these networks are presented in Table 5-4.

CHAPTER 5   HYPERPARAMETER TUNING

Table 5-4. *Hyperparameters of Autoencoders*

Network	Image	Hyperparameters	Description
Autoencoders		Encoder/decoder layers	The number of layers in the encoder and decoder.
		Latent dimension	The latent dimension signifies the size of the encoded representation.
		Learning rate	The learning rate controls the step size of the gradient descent update.
		Batch size	The batch size indicates the number of samples processed before the parameters of the model are updated.
		Dropout rate	The dropout rate denotes the fraction of the units to drop during training.

Experiments: Hyperparameter Tuning

This section presents an empirical analysis demonstrating the effect of hyperparameters on the performance of the model.

Problem: To classify the MNIST dataset

Data: The MNIST dataset consists of 60,000 training images and 10,000 test images of handwritten digits (0–9).

Architecture: Six different architectures (fully connected neural networks) are implemented with different numbers of hidden layers and numbers of neurons in each layer. The experiments also show the effect of variation in learning rate on the performance and the loss.

CHAPTER 5 HYPERPARAMETER TUNING

The models implemented in Listing 5-1 are as follows:

1. (512,)
2. (256,)
3. (128,)
4. (128, 64)
5. (128, 32)
6. (128, 16)

The individual plots of loss and accuracy for each model are shown in figures from Figure 5-5 to Figure 5-10. The variation of loss and accuracy with the number of epochs for different learning rates is plotted for the best model in Figure 5-11.

Listing 5-1. Hyperparameter tuning to classify the MNIST dataset

Code:

#1. The libraries *tensorflow* and specifically the *keras.models* and *keras.layers* are imported to design a sequential model having dense and flattened layers. We need to import the Adam optimizer from *tensorflow.keras.optimizers*

```
import tensorflow as tf
from tensorflow.keras.models import Sequential
from tensorflow.keras.layers import Dense, Flatten
from tensorflow.keras.optimizers import Adam
import matplotlib.pyplot as plt
import numpy as np
```

#2. We load the MNIST data set from *tensorflow.keras.datasets*, *mnist* and to get the train and test data we use *load_data()* function. Since the images are grayscale therefore the maximum value of a pixel is 255. If we divide every pixel by 255, we end up implementing Min-Max normalisation

```
mnist = tf.keras.datasets.mnist
(X_train, y_train), (X_test, y_test) = mnist.load_data()
X_train, X_test = X_train / 255.0, X_test / 255.0
```

#3. To compile the model, we use the compile function and set the parameters namely optimizer, loss, and metrics. Since it is a multiclass problem sparse categorical cross entropy is used as a loss function.

143

CHAPTER 5 HYPERPARAMETER TUNING

```python
def compile_and_train(model, lr=1e-3, epochs=10):
    model.compile(optimizer=Adam(learning_rate=lr),loss='sparse_
    categorical_crossentropy',metrics=['accuracy'])
    history = model.fit(X_train, y_train, epochs=epochs, validation_
    data=(X_test, y_test), verbose=0)
    return history
```

#4. Note that after compiling the model the output was saved in a variable called history. This is a dictionary from which training and validation accuracy are plotted.

```python
def plot_history(history, title):
    plt.figure(figsize=(12, 6))
    plt.plot(history.history['accuracy'], label='Train Accuracy')
    plt.plot(history.history['val_accuracy'], label='Validation Accuracy')
    plt.title(f'{title} Accuracy')
    plt.xlabel('Epochs')
    plt.ylabel('Accuracy')
    plt.legend()
    plt.show()
```

#5. The training and validation loss from history is plotted in the same way.

```python
    plt.figure(figsize=(12, 6))
    plt.plot(history.history['loss'], label='Train Loss')
    plt.plot(history.history['val_loss'], label='Validation Loss')
    plt.title(f'{title} Loss')
    plt.xlabel('Epochs')
    plt.ylabel('Loss')
    plt.legend()
    plt.show()
```

#6. The first model having a single hidden layer with 512 neurons is compiled and history is plotted.

```python
model_1 = Sequential([
Flatten(input_shape=(28, 28)),
Dense(512, activation='relu'),
Dense(10, activation='softmax')
])
```

```
history_1 = compile_and_train(model_1)
plot_history(history_1, 'Model [512]')
```
#7. The second model having a single hidden layer with 256 neurons is compiled and history is plotted.
```
model_2 = Sequential([
Flatten(input_shape=(28, 28)),
Dense(256, activation='relu'),
Dense(10, activation='softmax')
])
history_2 = compile_and_train(model_2)
plot_history(history_2, 'Model [256]')
```
#8. The third model having a single hidden layer with 128 neurons is compiled and history is plotted.
```
model_3 = Sequential([
Flatten(input_shape=(28, 28)),
Dense(128, activation='relu'),
Dense(10, activation='softmax')
])
history_3 = compile_and_train(model_3)
plot_history(history_3, 'Model [128]')
```
#9. The fourth model having two hidden layers with 128 and 64 neurons is compiled and history is plotted.
```
model_4 = Sequential([
Flatten(input_shape=(28, 28)),
Dense(128, activation='relu'),
Dense(64, activation='relu'),
Dense(10, activation='softmax')
])
history_4 = compile_and_train(model_4)
plot_history(history_4, 'Model [128, 64]')
```
#10. The fifth model having two hidden layers with 128 and 32 neurons is compiled and history is plotted.
```
model_5 = Sequential([
Flatten(input_shape=(28, 28)),
Dense(128, activation='relu'),
```

CHAPTER 5   HYPERPARAMETER TUNING

```
Dense(32, activation='relu'),
Dense(10, activation='softmax')
])
history_5 = compile_and_train(model_5)
plot_history(history_5, 'Model [128, 32]')
```
#11. The sixth model having two hidden layers with 128 and 16 neurons is compiled and history is plotted.
```
model_6 = Sequential([
Flatten(input_shape=(28, 28)),
Dense(128, activation='relu'),
Dense(16, activation='relu'),
Dense(10, activation='softmax')
])
history_6 = compile_and_train(model_6)
plot_history(history_6, 'Model [128, 16]')
```
#12. From history variable, we calculate the mean accuracy for all the models
```
mean_accuracies = {
    '[512]': np.mean(history_1.history['val_accuracy']),
    '[256]': np.mean(history_2.history['val_accuracy']),
    '[128]': np.mean(history_3.history['val_accuracy']),
    '[128, 64]': np.mean(history_4.history['val_accuracy']),
    '[128, 32]': np.mean(history_5.history['val_accuracy']),
    '[128, 16]': np.mean(history_6.history['val_accuracy'])
}
```
#13. Based on the above results the best architecture and model are printed
```
architecture = max(mean_accuracies, key=mean_accuracies.get)
print(f"Best architecture: {best_architecture} with mean accuracy: {mean_accuracies[best_architecture]:.4f}")
```
#14. We carry out an empirical analysis of the best model with different learning rates and plot the accuracy and loss curves.
```
learning_rates = [1e-4, 1e-3, 1e-2]
lr_histories = {}
for lr in learning_rates:
    model = create_model(eval(best_architecture))
```

```
    history = compile_and_train(model, lr=lr)
lr_histories[f'LR={lr}'] = history
# Plot accuracy and loss for different learning rates
plot_history(lr_histories, 'accuracy')
plot_history(lr_histories, 'loss')
```

Output:
Best architecture: [512] with mean accuracy: 0.9782

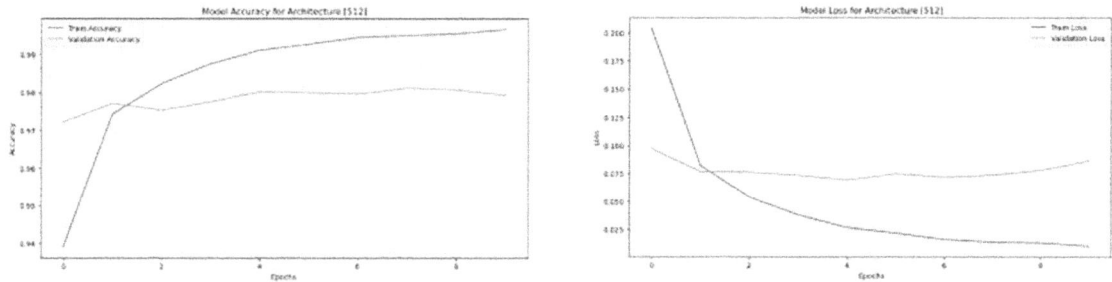

Figure 5-5. *Accuracy and loss curves for the architecture having a single hidden layer with 512 neurons*

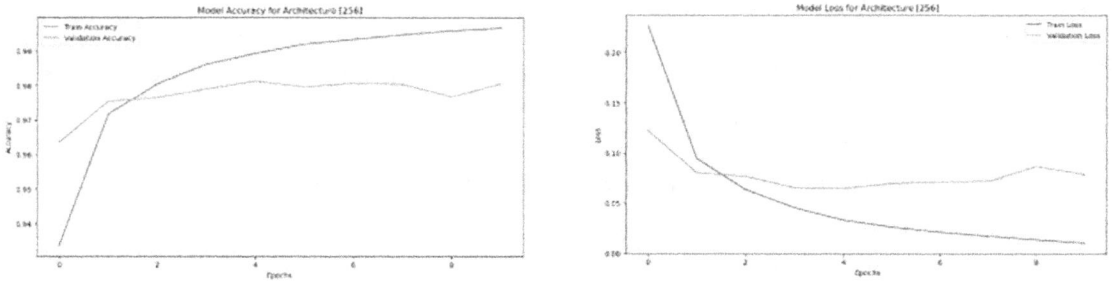

Figure 5-6. *Accuracy and loss curves for the architecture having a single hidden layer with 256 neurons*

CHAPTER 5 HYPERPARAMETER TUNING

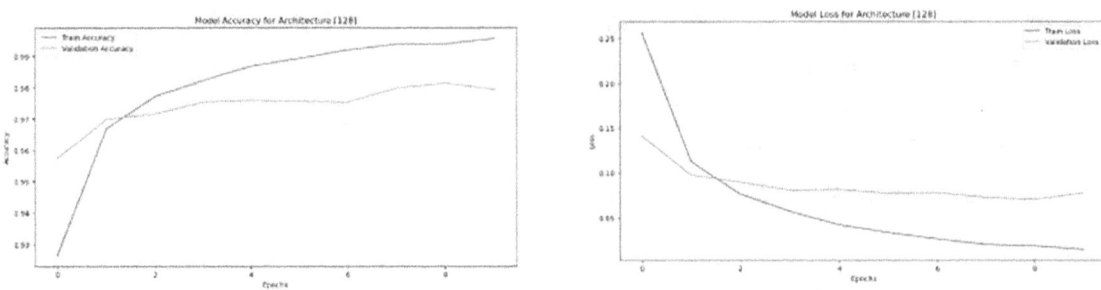

Figure 5-7. *Accuracy and loss curves for the architecture having a single hidden layer with 128 neurons*

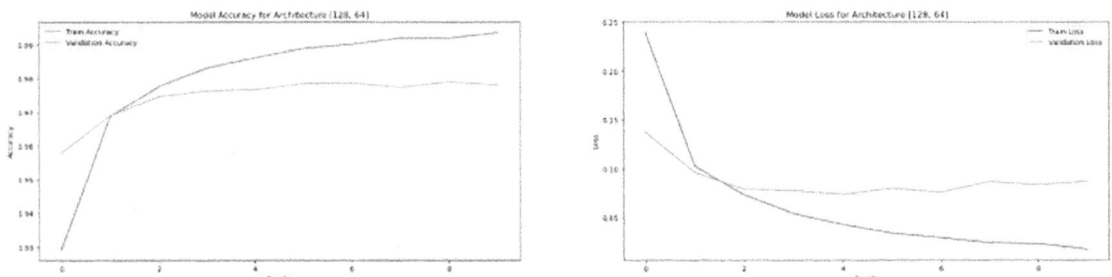

Figure 5-8. *Accuracy and loss curves for the architecture having two hidden layers with 128 and 64 neurons*

The following table (Table 5-5) shows the mean validation accuracy of six different architectures used to classify the MNIST dataset.

Table 5-5. *Mean Validation Accuracy of Six Different Architectures*

Architecture	Mean Validation Accuracy
(512,)	0.9782
(256,)	0.9769
(128,)	0.9732
(128, 64)	0.9738
(128, 32)	0.9737
(128, 16)	0.9725

CHAPTER 5 HYPERPARAMETER TUNING

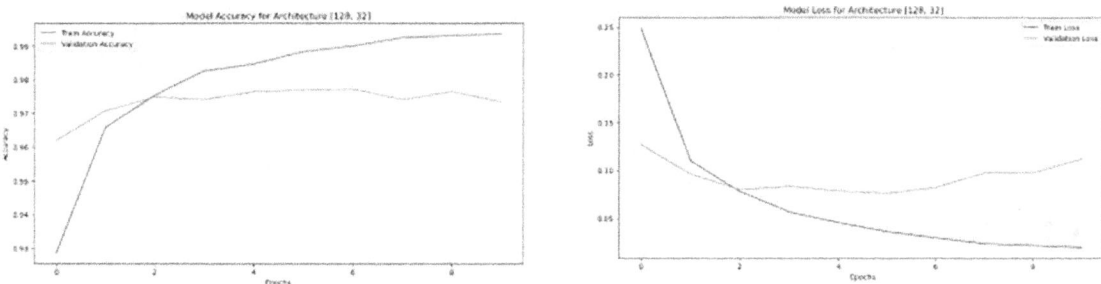

Figure 5-9. *Accuracy and loss curves for the architecture having two hidden layers with 128 and 32 neurons*

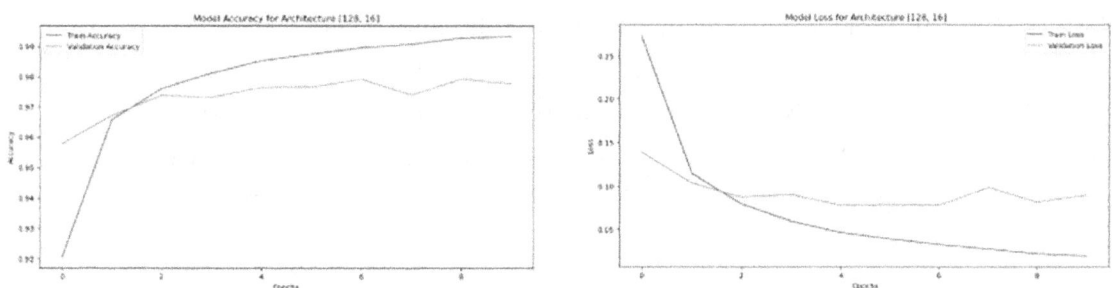

Figure 5-10. *Accuracy and loss curves for the architecture having two hidden layers with 128 and 16 neurons*

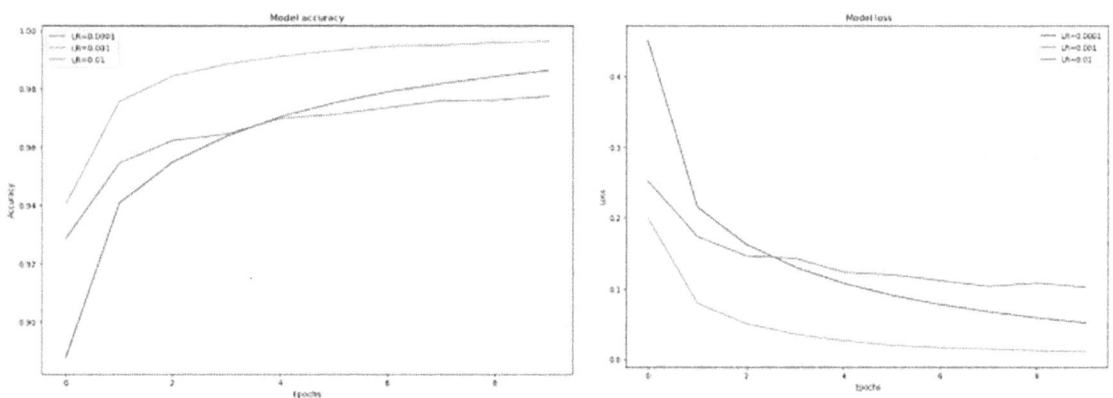

Figure 5-11. *Accuracy and loss curves for the best architecture (a single hidden layer with 512 neurons) with different learning rates*

CHAPTER 5   HYPERPARAMETER TUNING

Though the change in the accuracy is small, the performance of the model does depend on the number of hidden layers and the number of neurons in each hidden layer.

Conclusion

Deep Learning architectures are expected to perform well in the training as well as the test data. If the model does not perform well on the training data, we may need to revisit our assumptions regarding the data and the model that we are designing. If the model performs well on the training data but does not work well with the unseen data, then hyperparameter tuning may help us. This chapter discusses some important hyperparameters and their importance.

This discussion will also continue in the following chapters, as hyperparameter tuning is needed in CNNs, sequence models, and autoencoders as well. The reader is expected to attempt the exercise to get hold of the concept before moving forward.

Exercises

Multiple-Choice Questions

1. A Deep Neural Network must have at least one

    a. Output layer

    b. Input layer

    c. Hidden layer

    d. Dropout layer

2. If the number of layers in a Neural Network is too large, what problem might occur?

    a. Faster learning

    b. Overfitting

    c. Vanishing gradient

    d. Exploding gradient

CHAPTER 5 HYPERPARAMETER TUNING

3. Why is some depth required in a Neural Network?

    a. To increase the training time

    b. To decrease the complexity

    c. To extract the hierarchy of features

    d. To reduce training time

4. If the number of hidden layers is fewer and the number of neurons in each layer is high, what is generally preferred?

    a. Increase the number of layers and reduce the number of neurons in each layer.

    b. Decrease the number of layers and increase the number of neurons in each layer.

    c. Keep the number of layers and neurons the same.

    d. Increase both the number of layers and neurons.

5. What does the learning rate control in gradient descent?

    a. Batch size

    b. Step size of the gradient descent update

    c. Number of epochs

    d. Number of hidden layers

6. A lower learning rate results in which of the following?

    a. Faster learning

    b. More time to reach the optimal value

    c. Skipping the optimal value

    d. Overfitting

7. A higher learning rate may lead to which of the following?

    a. More time to reach the optimal value

    b. Skipping the optimal value in the loss landscape

    c. Reducing training time

    d. Better generalization

151

CHAPTER 5   HYPERPARAMETER TUNING

8. The batch size indicates

    a. The number of epochs

    b. The number of layers

    c. The number of samples processed before updating model parameters

    d. The learning rate

9. An epoch means

    a. The number of layers in the network

    b. The number of samples processed before updating model parameters

    c. The number of times the entire dataset is passed through the network

    d. The learning rate

10. The selection of the algorithm used to update weights affects the performance of the model. Which of the following are famous optimizers?

    a. SGD, RMSprop, Dropout

    b. Momentum, RMSprop, Dropout

    c. SGD, Momentum, RMSprop, Adam

    d. Adam, Dropout, SGD, RMSprop

11. According to the definition, an activation function in a Neural Network is

    a. The mapping of the input to the output via a linear transform function at each node

    b. The mapping of the input to the output via a nonlinear transform function at each node

    c. The mapping of the output to the input via a nonlinear transform function at each node

    d. The mapping of the output to the input via a linear transform function at each node

CHAPTER 5 HYPERPARAMETER TUNING

12. Regularization trades a marginal decrease in training accuracy for which of the following?

    a. An increase in training speed

    b. An increase in overfitting

    c. An increase in generalizability

    d. An increase in batch size

13. The dropout rate denotes which of the following?

    a. The fraction of the units to drop during training

    b. The fraction of the units to add during training

    c. The learning rate of the network

    d. The number of epochs

14. Which of the following increases the generalizability of the model?

    a. Dropout

    b. Lower learning rate

    c. High learning rate

    d. None of the above

15. Which of the following may be considered for decreasing the variance of the model?

    a. Dropout

    b. Large training set

    c. Regularization

    d. All of the above

CHAPTER 5　HYPERPARAMETER TUNING

Experiments

I. The CIFAR dataset (https://www.cs.toronto.edu/~kriz/cifar.html) has 60,000 images belonging to 10 classes. Each class has 6000 images. The dataset is divided into two parts, train and test, having 50,000 and 10,000 images, respectively.

   Download the dataset and design a fully connected network having two hidden layers to classify this dataset. You can find the number of neurons in each hidden layer by carrying out empirical analysis. Train your network and report the results.

   Carry out the following tasks and report the results, as expressively as you can.

   1. Retrain the network with the following optimizers and analyze the performance and the effect on the loss curve:

      a. Adam optimizer

      b. RMSprop

      c. Momentum

   2. Use a dropout layer to reduce the variance of the model.

   3. Find the effect of change in the learning rate on the performance of the model.

   4. Use regularization to see if the model gives better results with the test data.

   5. Does changing the activation function in each layer affect the smoothness of the loss curve?

II. Now explore the STL dataset (https://cs.stanford.edu/~acoates/stl10/#:~:text=The%20STL%2D10%20dataset%20is,dataset%20but%20with%20some%20modifications) and perform the tasks stated in the above question again.

References

[1] Termanini, R. (2020). Synthesizing DNA-encoded data. In Elsevier eBooks (pp. 173–224). https://doi.org/10.1016/b978-0-12-823295-8.00007-0

[2] Murel, J., PhD, & Kavlakoglu, E. (2024, September 2). Regularization. What is regularization? https://www.ibm.com/topics/regularization

CHAPTER 6

Convolutional Neural Networks: I

Hubel and Wiesel proposed that the pattern recognition tasks in monkeys and cats use two types of cells, one of which has a larger receptive field. The output of this field does not depend on the location of the edges in the field. This inspired Kunihiko Fukushima to introduce neo-cognition, which in turn inspired convolutional and downsampling layers in Neural Networks, called Convolutional Neural Networks (CNNs). Backpropagation was used in the CNNs by Yann LeCun, a French computer scientist and the recipient of the prestigious Turing Award. LeNet, the first CNN, could recognize handwritten digits. Ignored initially, the CNNs got their due share in 2012, with the advent of AlexNet. CNNs have been successfully applied to image classification, object detection, and disease prediction and even in digital arts. They have shown better performance compared with the existing Neural Networks and are being extensively used in numerous disciplines.

Let us begin our discussion with the comparison of the Multi-layer Perceptron (MLP) and Convolutional Neural Networks. The former has already been discussed in the previous chapters. The MLP has an input layer, an output layer, and at least one hidden layer. A neuron in a hidden layer receives inputs, multiplies them with weights, and adds biases to the product. The result is then fed to an activation function. The output of this neuron may act as an input to another neuron. The forward pass is followed by a backward pass. Convolutional Neural Networks follow the same principle but are specifically designed for images. They use the modified convolution operator to extract the feature maps. These models have many types of layers like the convolutional layer, which helps in finding the feature maps; the pooling layer, which helps in downsampling; the activation layer; and fully connected layers. This chapter discusses the various types of layers in CNN and explains their need. It may be noted that only some of these layers have hyperparameters and learn the weights. Figure 6-1 summarizes the discussion.

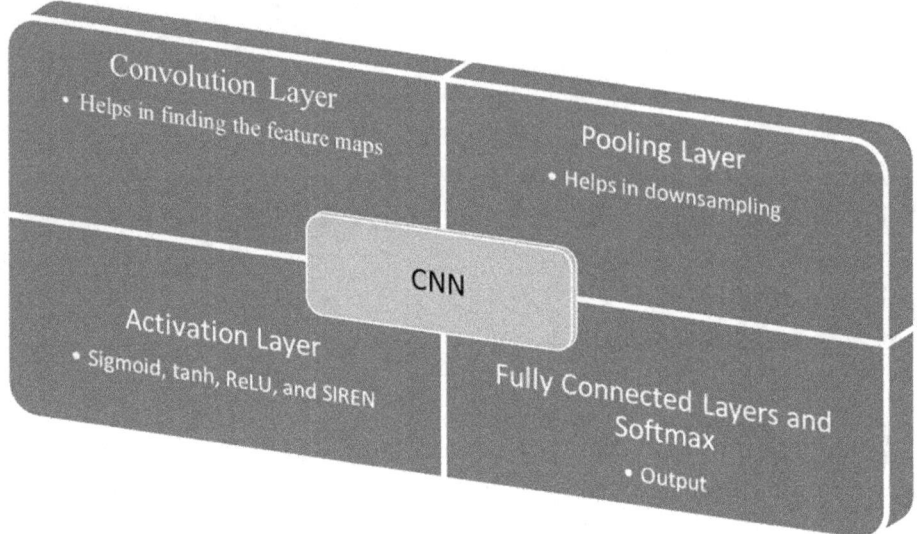

Figure 6-1. *Components of a CNN*

The reader may take note of the fact that there are notable differences between MLP and CNN. In MLP the neurons of a layer ***do not share connections***, whereas, in the case of CNN, they do. This is important. For example, consider a fully connected neural network that takes a grayscale image that has dimensions 300 × 300 × 1 as input, has 50 neurons in the output layer, and 100 neurons in hidden layers; then the learnable weights will be 100 × 300 × 300 + 50 × 100 = 9000000 + 5000 = 9005000. Along with these, there will be 150 biases, thus resulting in 90050150 learnable parameters. The CNN uses filters, explained in the following sections. If a 10 × 10 filter is used for extracting the relevant features, then there will be 101 (one bias) learnable parameters. Even if there are ten such filters, the number of learnable parameters will be 1010. Likewise, there will be some learnable parameters in the output-hidden layers. The total number of ***learnable parameters is still much fewer compared*** with a fully connected MLP. This chapter explores the idea of filters and the need for many filters.

So, in a fully connected MLP, each neuron is connected to all the neurons in the previous layer, whereas in the case of a CNN, some neurons are connected to only a portion of the previous layer. This gives rise to the concept of shared weights. The earlier layers of a CNN are expected to find ***low-level features*** like edges, and the later layers are expected to find ***high-level features***. Moreover, this connection of a filter to a small portion of the previous layer results in a type of regularization. Moreover, CNNs are

CHAPTER 6   CONVOLUTIONAL NEURAL NETWORKS: I

translation and rotation invariant. This makes sense as in a recognition task; we would like to find an object irrespective of its position in the image. Likewise, even if the object is rotated, the model should be able to find the object.

Tip   CNN vs. MLP

- CNNs take into account spatial correlation; MLPs do not.
- CNNs have fewer learnable parameters.
- CNNs are translation and rotation invariant.

This chapter introduces the components of CNN. The implementation of these units from scratch will not only help the reader in understanding the working of the component but will also empower them to make changes in the component as and when required. The chapter has been organized as follows. Section "Convolutional Layer" discusses the convolution operator. The next section, "Implementing Convolution," presents the implementation of the convolution operator and discusses its importance. The next section, "Padding," discusses padding. The next section, "Stride and Other Layers," explains stride and discusses other layers, and the next section, "Importance of Kernels," explains the importance of convolution. This is followed by a brief introduction to LeNet, the first CNN. The last section concludes.

Convolutional Layer

In this layer, filters ***extract features from the input tensor*** and ***create a feature map***. To understand this, consider a grayscale image, which can be represented as a matrix having values ranging between 0 and 255. A 2D kernel is a filter that is expected to find the prominent features of the given image. The convolution of the given image and the kernel give the output. For example, if the input is a matrix of dimension 8 × 8 and a kernel has dimension 3 × 3, we initially place the kernel at the top-left corner of the matrix and find the sum of the products of the corresponding elements. Consider, for example, the input image and the kernel shown in Figure 6-2. The result of the initial convolution operation would be 120.

CHAPTER 6   CONVOLUTIONAL NEURAL NETWORKS: I

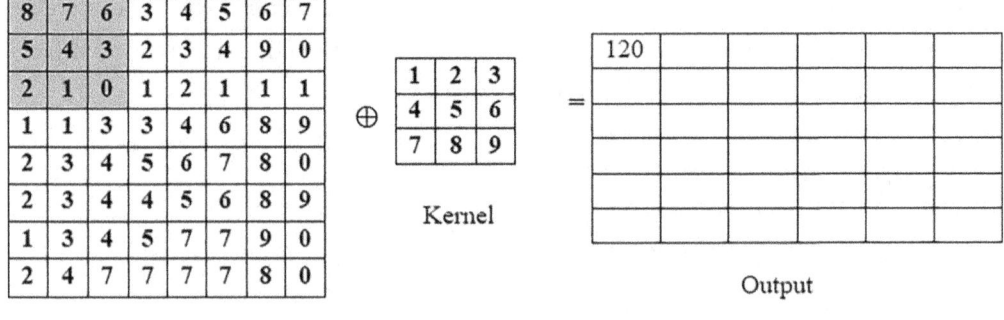

Figure 6-2. *The convolution operation*

$$\text{Result} = 8\times1+7\times2+6\times3+5\times4+4\times5+3\times6+2\times7+1\times8+0\times9$$

$$= 8+14+18+20+20+18+14+8+0$$

$$= 120$$

Now, let's shift the kernel one step to the right and find the sum of products again. The amount by which the kernel moves in a unit of time is called **stride**. Note that there will be six such products for the first row (Figure 6-3). Likewise, there will be six such rows. That is, this operation will result in six values per row and six such rows. The output will, therefore, be a 6 × 6 matrix.

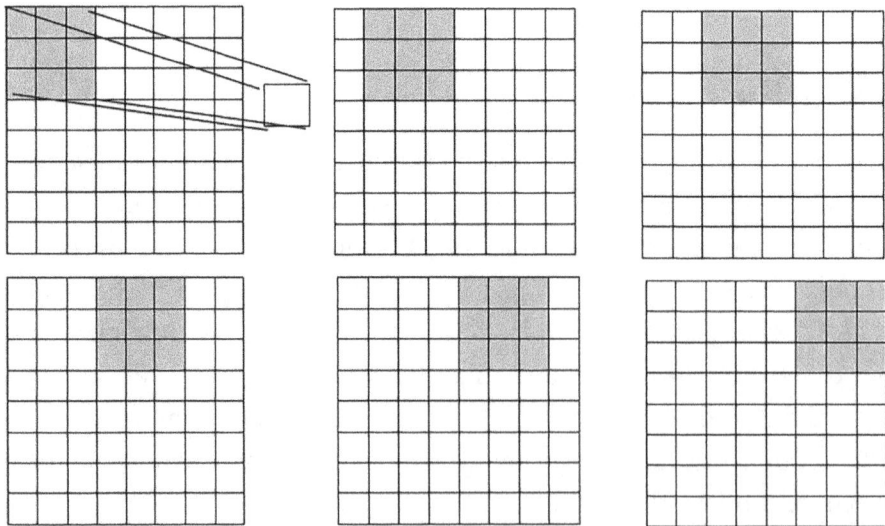

Figure 6-3. *With stride = 1, kernel size = 3, and input size = 8, there will be six outputs, for each row*

CHAPTER 6 CONVOLUTIONAL NEURAL NETWORKS: I

In general, for input size $n \times n$, kernel size $k \times k$, and stride s, the size of the output will be

$$m = \frac{n-k}{s} + 1.$$

Having seen the working of this layer, let us now move to the implementation of this operation. The reader may note that this convolution operation is not the same as that in signal processing.

Implementing Convolution

To understand the advantages of this operation, consider the following kernels. The convolution of the first kernel (kernel-1) with the image results in a feature map in which the horizontal lines can be seen. Likewise, the convolution of kernel-2 with the image results in a feature map with vertical lines (Figure 6-4).

-1	-1	-1
0	0	0
1	1	1

Kernel-1

-1	0	1
-1	0	1
-1	0	1

Kernel-2

Figure 6-4. *Kernels that extract horizontal and vertical lines in an image*

The following code implements convolution with stride = 1. The first step imports the required modules. The second step reads the input image, and the third step converts it into a grayscale. Step 4 creates the above kernel and implements convolution. The fifth step applies convolution to the image.

CHAPTER 6   CONVOLUTIONAL NEURAL NETWORKS: I

Step 1: Import Matplotlib and NumPy.
Code:

```
from matplotlib import pyplot as plt
import numpy as np
```

Step 2: Read an image.
Code:

```
arr=plt.imread('Juggie.jpg')#The image can be found in web resources
print(arr.shape)
plt.imshow(arr)
```

Output:

(281, 180, 3)

Step 3: Convert the colored image to grayscale.

Code:

```
def rgb2gray(rgb):
    r, g, b = rgb[:,:,0], rgb[:,:,1], rgb[:,:,2]
    gray = 0.2989 * r + 0.5870 * g + 0.1140 * b
    return gray
```

Step 4(a): Create the kernel.

Code:

```
Kernel=[[2,0,-2],[2,0,-2],[2,0,-2]]
Kernel=np.array(Kernel)
```

Step 4(b): Apply convolution to the image.

Code:

```
def conv(Image, Kernel):
    n=Image.shape[0]
    m=Image.shape[1]
    k=Kernel.shape[0]
    new_image=np.zeros((n-k+1, m-k+1))
    for i in range(n-k+1):
        for j in range(m-k+1):
            arr1=Image[i:i+k, j:j+k]
            ans=np.sum(arr1*Kernel)
            new_image[i,j]=ans
    return new_image
```

Step 5: Apply convolution to the image.

Code:

```
result=conv(arr_gray,Kernel )
plt.imshow(result)
```

CHAPTER 6 CONVOLUTIONAL NEURAL NETWORKS: I

Output:

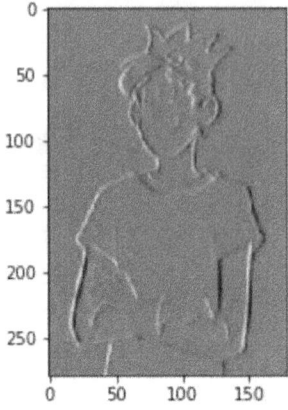

The reader should also run the above code with the following kernel

```
Kernel=[[2,2,2],[0,0,0],[-2,-2,-2]]
Kernel=np.array(Kernel)
```

and observe the output. The expected output should be like the one shown in the following figure.

Output:

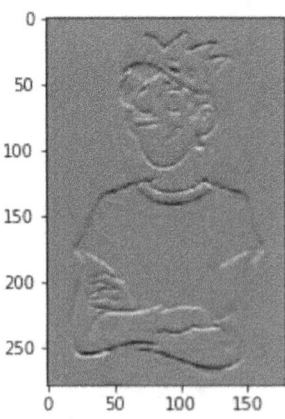

Note that in the first output, the vertical lines are prominent and in the second, the horizontal lines are prominent. This is what a kernel is expected to do. We may have a kernel that finds the vertical lines, another may find the horizontal lines, and yet another finds the inclined ones. Having **more than one kernel** will yield important features of a given image.

Now, take a pause and think, What if the weights of a kernel could be **learned**? That would be wonderful! This will **allow the layer to find requisite features from a given image** and may help in tasks like classification and so on. The importance of these kernels is explained in section "Importance of Kernels."

Padding

At times the convolution operation cannot traverse the whole image. For example, if the size of the input image is 5 × 5, the size of the kernel is 3 × 3, and the stride is 3, there will be a problem. This can be resolved by padding the image with zeros. For example, consider an image of size 5 × 5; padding of p = 2 will result in an image of size 9 × 9, as shown in Figure 6-5.

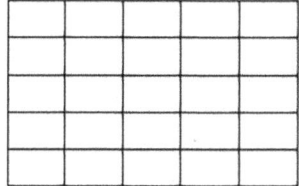
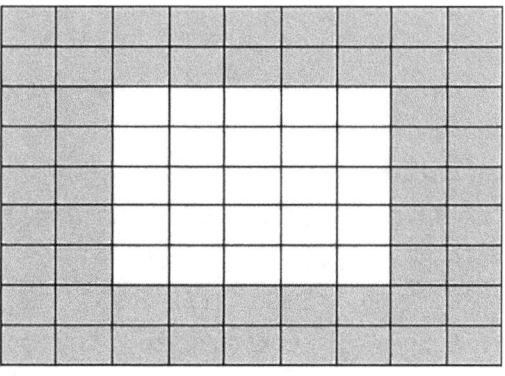

Figure 6-5. *Input image is padded with zeros (p = 2).*

The following code implements padding. The function takes the image (dimensions: $n \times m$) and the value of p as a parameter and produces an image of dimensions

$$(n+2p)\times(m+2p).$$

Here, a random array is created, and padding of p = 2 is applied to the so-formed image.

Code:

```python
#Create random array
arr=np.random.randint(0,255,(30,30))
plt.imshow(arr)
```

Output:

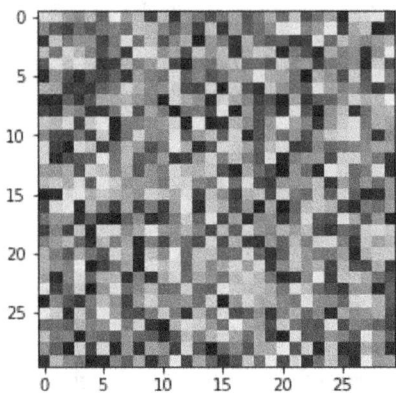

Code:

```python
#Define function
def pad(img, p):
    arr1=np.zeros((p,img.shape[1]+2*p))
    arr2=np.zeros((img.shape[0],p))
    arr_temp=np.hstack((arr2,img))
    arr_temp=np.hstack((arr_temp,arr2))
    arr_temp=np.vstack((arr1,arr_temp))
    arr_temp=np.vstack((arr_temp,arr1))
    return(arr_temp)
#Pass the array in the function
img1=pad(arr,2)
print(img1.shape)
plt.imshow(img1)
```

CHAPTER 6 CONVOLUTIONAL NEURAL NETWORKS: I

Output:

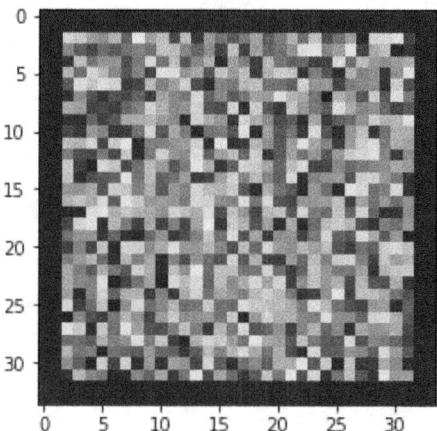

It may also be noted that in the case of padding, the dimensions of the output field change. The dimensions of the output, if the padding is P, kernel size is F, and stride is S, can be calculated using the following formulas:

$$W = \frac{(W-F+2P)}{S}+1$$

$$H = \frac{(H-F+2P)}{S}+1$$

Stride and Other Layers

Having seen the implementation of the convolutional layer and padding, let us now move to pooling. However, before that let's have a quick look at the idea of stride.

Stride

The number of steps by which the kernel moves forward, in a unit of time, is referred to as stride. In the above discussion and implementation, the stride was taken as 1. Figure 6-6 considers the value of stride as 2.

CHAPTER 6   CONVOLUTIONAL NEURAL NETWORKS: I

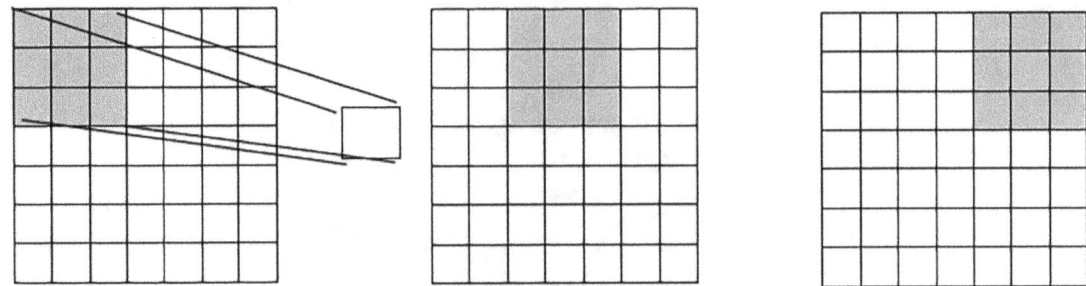

Figure 6-6. *Stride = 2*

Note that the more the value of s, the lesser the size of the output image. In general, with stride s, the size of the output will be given by a formula given below.

Pooling

Generally, a pooling layer is inserted between two consecutive convolutional layers. This **helps in reducing the size of the existing representation**. This size reduction is important because of two reasons: firstly, in reducing the number of parameters, and, secondly, in controlling the overfitting of the model. As per the literature review, pooling can be done by (a) taking out the maximum of the given window or (b) by taking the average or (c) taking the sum. So, if the size of the window is $W \times H \times D$ and the spatial extent of the pooling layer is F, then the size of the output layer is given by

$$W = \frac{(W-F)}{S} + 1$$

$$H = \frac{(H-F)}{S} + 1$$

while the depth, that is, D, remains the same.

It may also be noted that max pooling is more popular as compared with all other types of pooling. The following code carries out the pooling of a given image. The reader is expected to observe the images after and before applying the pooling operation and figure out why an object can be recognized even after applying pooling.

Code:

```
def pooling(image,E,S):
    n = image.shape[0]
    m = image.shape[1]
    new_arr = np.zeros( (((n-E)//S)+1,((m-E)//S)+1))
    p=0
    k=0
    for i in range(((n-E)//S)+1):
        k=0
        for j in range(((m-E)//S)+1):
            arr = image[i:i+E,j:j+E]
            ans = np.max(arr)
            new_arr[p,k] = ans
            k+=1
        p+=1
    return new_arr
```

Note that replacing "ans = np.max(arr)" with "ans = np.sum(arr)" will result in sum pooling and "np.mean" will result in average pooling.

Normalization

The concept of the normalization layer was introduced for mimicking the inhibition scheme of our brain. However, they have not proved to be much of a benefit; thus, they are not much in use. There are various types of normalization techniques, some of which are as follows:

- Local response normalization layer (same map)
- Local response normalization layer (across maps)
- Local contrast normalization layer

The interested readers may refer to the References given at the end of this Chapter for more details.

CHAPTER 6 CONVOLUTIONAL NEURAL NETWORKS: I

Fully Connected Layer

As the name suggests, in this layer, each neuron in a layer is connected to each neuron of the previous layer. Thus, we can say that they behave as a normal Neural Network. The topic has already been discussed in the previous chapters.

Importance of Kernels

Having seen the basics of each type of layer, let us move back to the importance of kernels, which are the most important components of a CNN. Consider Pattern 1. The kernel shown in Figure 6-7 finds the horizontal lines in the pattern. When the convolution operation is applied to the pattern with this kernel, the picture shown in the output that follows is produced. Note that if the kernel is slightly changed, a slightly different output is produced. The outputs show the regions where the line starts and ends. The reader is expected to run the code, which follows, and identify the intensities of the lighter and the darker lines in the output. As a matter of fact, they represent the positive and negative edges.

Pattern 1:

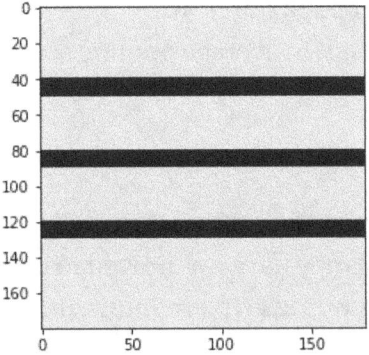

2	2	2
0	0	0
-2	-2	-2

-2	-2	-2
0	0	0
2	2	2

Kernel 1   **Kernel 2**

Figure 6-7. *Kernel 1 and Kernel 2 can identify horizontal lines*

Code:

```
Kernel_horz1=np.array([[2,2,2],[0,0,0],[-2,-2,-2]])
result1=conv_stride(pattern1,Kernel_horz1,1)
plt.imshow(result1)
```
Output:

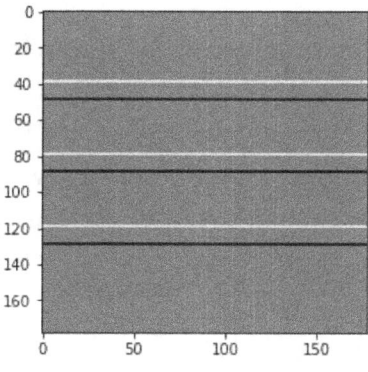

2	2	2
0	0	0
-2	-2	-2
-2	-2	-2
0	0	0
2	2	2

CHAPTER 6 CONVOLUTIONAL NEURAL NETWORKS: I

Code:

```
Kernel_horz1=np.array([[-2,-2,-2],[0,0,0],[2,2,2]])
result1=conv_stride(pattern1,Kernel_horz1,1)
plt.imshow(result1)
```

Output:

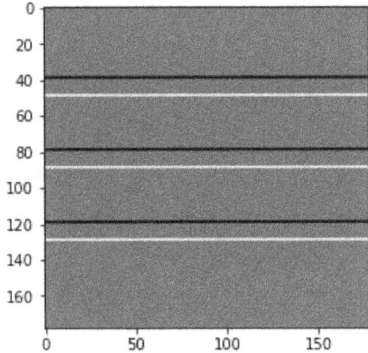

Now, consider Pattern 2. The kernel shown in Figure 6-8 finds the vertical lines in the pattern. When the convolution operation is applied to the pattern with this kernel, the picture shown in the output is produced. Note that if the kernel is slightly changed, a slightly different output is produced. The outputs show the regions where the line starts and ends. The reader is expected to run the code, which follows, and identify the intensities of the lighter and the darker lines in the output. Again, they represent the positive and negative edges.

Pattern 2:

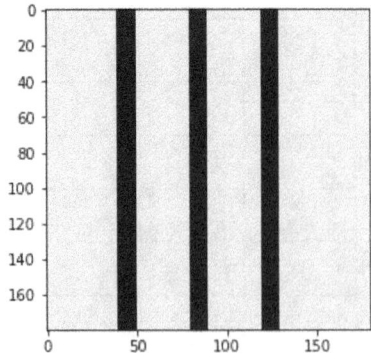

CHAPTER 6 CONVOLUTIONAL NEURAL NETWORKS: I

2	0	-2
2	0	-2
2	0	-2

-2	0	2
-2	0	2
-2	0	2

Kernel 3 **Kernel 4**

Figure 6-8. Kernel 3 and Kernel 4 can identify vertical lines

---

Code:

```
Kernel_vert1=np.array([[2,0,-2],[2,0,-2],[2,-0,-2]])
result3=conv_stride(pattern2,Kernel_vert1,1)
plt.imshow(result3)
```

Output:

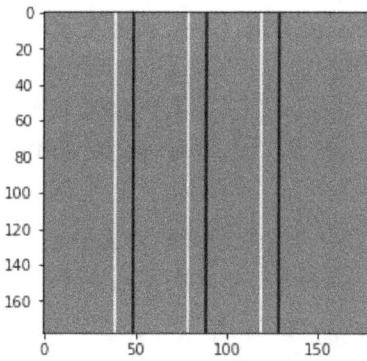

2	0	-2
2	0	-2
2	0	-2
-2	0	2
-2	0	2
-2	0	2

CHAPTER 6 CONVOLUTIONAL NEURAL NETWORKS: I

Code:

```
Kernel_vert2=np.array([[-2,0,2],[-2,0,2],[-2,-0,2]])
result4=conv_stride(pattern2,Kernel_vert2,1)
plt.imshow(result4)
```

Output:

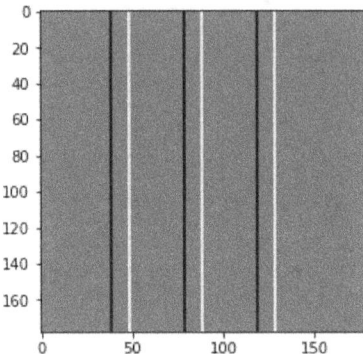

Tip Note that if the kernel capable of finding the horizontal lines is applied to the picture containing the vertical lines (or vice versa), nothing is produced.

Code:

```
result5=conv_stride(pattern2,Kernel_horz1,1)
plt.imshow(result5)
```

Output:

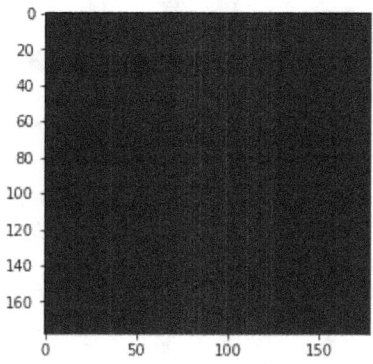

Code:

```
result6=conv_stride(pattern1,Kernel_vert1,1)
plt.imshow(result6)
```

Output:

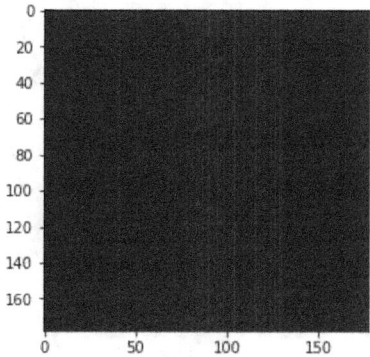

The reader is expected to apply the above kernels in the following pattern (Pattern 3) and observe the results.

CHAPTER 6 CONVOLUTIONAL NEURAL NETWORKS: I

Pattern 3:

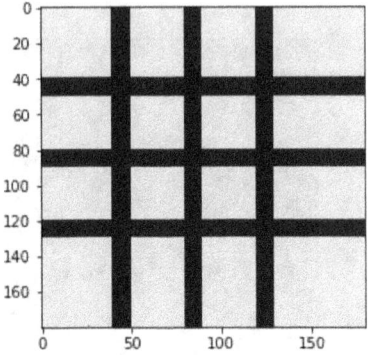

Code:

```
result7=conv_stride(pattern3,Kernel_horz1,1)
plt.imshow(result7)
```

Output:

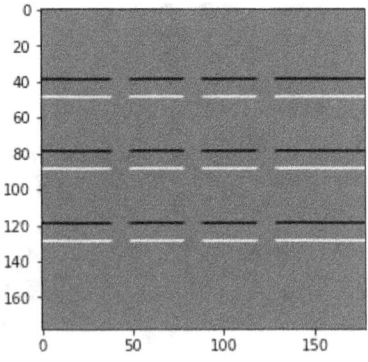

Code:

```
result8=conv_stride(pattern3,Kernel_vert1,1)
plt.imshow(result8)
```

Output:

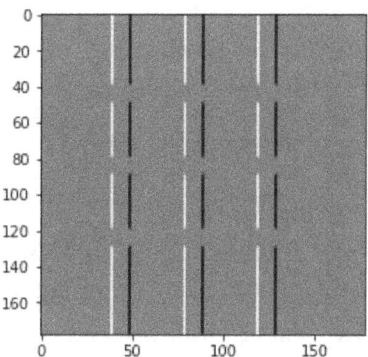

Note that the first two kernels can find horizontal lines and the next two kernels can find vertical lines. Likewise, some kernels can find diagonals, and so on. If the information obtained by the application of some kernels is combined, the texture information regarding the input can be retrieved. This is what the convolution operation does. Moreover, in the above discussion, the kernels were chosen. In CNN, kernels are learned, which makes the output quite informative. This output can, hence, extract the information regarding the texture of a given image in a better way. The above discussion will help the reader appreciate the need for multiple kernels.

Having studied various types of layers in a CNN, let us consider one of the simplest CNNs called LeNet, which is capable of recognizing handwritten digits.

Architecture of LeNet

LeNet was introduced in 1998 by LeCun et al. in the paper titled "Gradient-Based Learning Applied to Document Recognition." The original paper described the LeNet 5 architecture, which had the following layers:

- Convolution: 6 layers having kernel size = 5, stride = 1, output = 28 × 28
- Sub-sampling: Average pooling output = 14 × 14
- Convolution: 16 layers having kernel size = 5, stride = 1

CHAPTER 6   CONVOLUTIONAL NEURAL NETWORKS: I

- Sub-sampling: Average pooling output
- Convolution: 120 layers having kernel size = 5, stride = 1
- Flatten layer
- Dense layer: 84 neurons, activation = tanh
- Dense layer: 10 neurons, activation = softmax

The original paper describing LeNet can be found at http://vision.stanford.edu/cs598_spring07/papers/Lecun98.pdf. Note that by stacking alternate layers of convolution and pooling followed by some fully connected layers, some interesting architectures can be crafted. The next chapter presents the Keras implementation of each of the layers discussed above and discusses the design of a sequential model. However, the reader may refer to the following code, which implements LeNet and presents its application with the popular MNIST dataset containing images of handwritten digits.

Code:

```
#Importing Libraries
import tensorflow as tf
from tensorflow.keras import datasets, layers, models
import matplotlib.pyplot as plt
#Split the data into train and test set
(X_train, y_train), (X_test, y_test) = datasets.mnist.load_data()
#Normalization
X_train, X_test = X_train / 255.0, X_test / 255.0
#Displaying the shape of the train and the test data
print(X_train.shape, X_test.shape)
#Convention: (number of samples, x, y, z)
X_train = X_train.reshape(X_train.shape[0], 28, 28, 1)
X_test = X_test.reshape(X_test.shape[0], 28, 28, 1)
#Displaying new shapes
print(X_train.shape, X_test.shape)
#Developing model
LeNet = models.Sequential()
LeNet.add(layers.Conv2D(6, (5, 5), activation='relu', input_shape=
(28, 28, 1)))
```

```
LeNet.add(layers.MaxPooling2D((2, 2)))
LeNet.add(layers.Conv2D(16, (5, 5), activation='relu'))
LeNet.add(layers.MaxPooling2D((2, 2)))
LeNet.add(layers.Flatten())
LeNet.add(layers.Dense(120, activation='relu'))
LeNet.add(layers.Dense(84, activation='relu'))
LeNet.add(layers.Dense(10, activation='softmax'))
LeNet.compile(optimizer='adam',loss='sparse_categorical_crossentropy',
metrics=['accuracy'])
LeNet.summary()
#For observing the variation in loss and performance with iteration
history = LeNet.fit(X_train, y_train, epochs=25, validation_data=
(X_test, y_test))
#Plotting Loss and Accuracy of Training and Validation
plt.plot(history.history['loss'], label='Training Loss')
plt.plot(history.history['val_loss'], label='Validation Loss')
plt.title('Training and Validation Loss')
plt.xlabel('Epochs')
plt.ylabel('Loss')
plt.legend()
plt.show()
plt.plot(history.history['accuracy'], label='Training Accuracy')
plt.plot(history.history['val_accuracy'], label='Validation Accuracy')
plt.title('Training and Validation Accuracy')
plt.xlabel('Epochs')
plt.ylabel('Accuracy')
plt.legend()
plt.show()
```

CHAPTER 6   CONVOLUTIONAL NEURAL NETWORKS: I

Output:

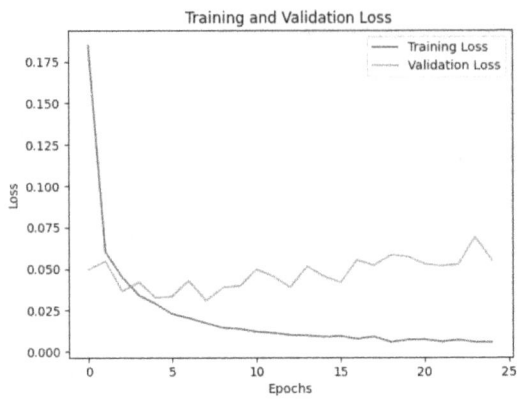

The next chapter revisits LeNet and compares it with AlexNet. It also discusses why this architecture works wonders with handwritten digits but does not perform well with complex images.

Conclusion

The previous chapters of this book discussed MLP. There are two major problems with these methods:

   i. In these networks the number of connections is huge; therefore, the learning requires many inputs and takes time.

   ii. This model does not take into account the spatial correlation.

CHAPTER 6   CONVOLUTIONAL NEURAL NETWORKS: I

This chapter introduced the components of Convolutional Neural Networks. It discusses the importance of convolution and presents the implementations of the pooling layer, the convolutional layer, etc. Also, the convolution operation explained is slightly different from the mathematical convolution.

The reader should be able to implement the layers from scratch using NumPy after reading this chapter. Also, the reader is expected to appreciate the importance of multiple kernels in CNN. However, one need not implement everything from scratch; Keras provides the implementations of all the layers. The next chapter ***introduces Keras*** and explains the implementation of layers using Keras. The chapter also introduces some of the most ***important CNNs and their implementations***. It will empower you with the most powerful weapons to fight the problems of image analysis. To conclude, we started with neurocognition. The following image (Figure 6-9) of neurocognition has been generated by AI (`https://gencraft.com/generate`) and uses CNN.

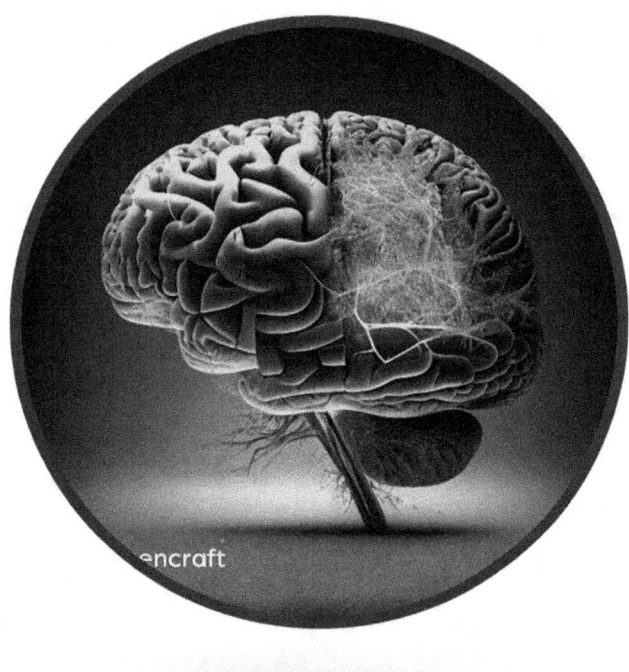

Figure 6-9. *Image of neurocognition generated by* `https://gencraft.com/generate`

Before proceeding any further, let's test our understanding.

CHAPTER 6 CONVOLUTIONAL NEURAL NETWORKS: I

Exercises

Multiple-Choice Questions

1. CNN is generally used for which of the following?

    a. Images

    b. Text

    c. Sound

    d. None of the above

2. Which of the following tasks can be accomplished using CNN?

    a. Image classification

    b. Image detection

    c. Segmentation

    d. All of the above

3. For classifying sounds, which of the following can be used?

    a. CNN

    b. RNN

    c. MLP

    d. All of the above

4. Convolution uses

    a. Shared weights

    b. Neurological analogy

    c. Both

    d. None of the above

5. How many kernels can a convolutional layer have?

    a. Only one

    b. More than one

c. Cannot say

d. None

6. Which of the following reduces the size of the output?

    a. Pooling

    b. Spooling

    c. Schooling

    d. Cooling

7. Generally, which of the following are used in pooling?

    a. Maximum

    b. Average

    c. Sum

    d. Any of the above

8. Can CNN have multiple convolutional layers?

    a. Yes

    b. No

9. The fully connected layers (s) with respect to a CNN

    a. Are generally the last layers

    b. Are generally placed at the beginning of the network

    c. Are middle layers

    d. None of the above

10. Which of the following is not a layer of a CNN?

    a. Convolution

    b. Fully connected

    c. LTU

    d. Pooling

CHAPTER 6   CONVOLUTIONAL NEURAL NETWORKS: I

Numerical

1. If the size of an image is 20 × 20, that of the kernel is 5 × 5, and stride = 1, what should be the value of p so that the size of the output image is the same as that of the input?
2. If the size of an image is 20 × 20, that of the kernel is 5 × 5, and stride = 2, what should be the size of the output image if p = 1 and p = 2?
3. If the size of an image is 20 × 20, that of the kernel is 5 × 5, and stride = 1, what should be the size of the output if p = 0?
4. In the above case, if s = 2, what should be the size of the output image?
5. Find the size of the kernel that produces an image of size 20, for an input image of size 20, if the value of p is 2 and s = 1.

Applications

1. State the filters for finding the horizontal and vertical lines.
2. Suggest a filter for finding diagonals in an image.
3. What happens if the rows containing 2s are swapped with the row containing -2s in the following kernel?

Kernel 1

4. Can you find both horizontal and vertical lines using a single filter?
5. Can the above task be accomplished using two filters?
6. You are required to classify the images of oranges and apples. The images are of size 100 × 100. Suggest a Multi-layer Perceptron to accomplish this task. Also implement the network using Keras.
7. Accomplish the above task using a CNN. (The reader may attempt this after reading the next chapter.)
8. Compare the number of learnable weights in the above two structures.

CHAPTER 7

Convolutional Neural Network: II

The last chapter discussed the units of a Convolutional Neural Network and introduced LeNet. This chapter takes the discussion forward and presents an overview and implementation of some of the famous CNN architectures like LeNet, AlexNet, and Google LeNet (Inception Net). The simplicity of LeNet gives a good idea of how things work in CNN. However, to classify complex images and to accomplish advanced image analysis tasks, we need deep, more complex structures. The advancements in the 2010s were aimed at handing the problems in the then-popular CNNs and gave us the architectures that have since become immensely important for all image-related tasks: both supervised and unsupervised.

In the last chapter, the CNN layers were implemented from scratch, which is practically not required. In this chapter, the sequential model of Keras is explained, and requisite examples are presented to help the reader implement basic CNN. This chapter also presents a brief overview of some of the most important layers in Keras.

This chapter has been organized as follows. Section "Sequential Model" discusses the sequential model; section "Keras Layers" presents an overview of ***keras.layers***. Section "MNIST Dataset Classification Using LeNet: Prerequisite" implements an MNIST classifier. The next three sections discuss LeNet, AlexNet, and other important CNN models, and the last section concludes. This chapter forms the basis of the following chapters and will help you accomplish tasks like object detection and segmentation.

CHAPTER 7 CONVOLUTIONAL NEURAL NETWORK: II

Sequential Model

The sequential model comes to our rescue when we need to stack layers in a model, which takes a tensor as input and produces a tensor (TITO: Tensor Input Tensor Output). However, if the model has multiple inputs, then the sequential model is not used. Also, in the case of a model having multiple outputs or in the case of nonlinear models, they are not used. The following imports are required for building the model.

Code:

```
import tensorflow as tf
from tensorflow import Keras
from tensorflow.keras import layers
```

Creating the Model

You can create a sequential model by passing a list of layers in the **keras.Sequential** method. For example, the following code creates a sequential model with three layers. The input to the model is a 10 × 10 tensor. The rest of the arguments of the **layers.Dense** are explained in the sections that follow.

Code:

```
model = keras.Sequential(
    [
        layers.Dense(5, activation="relu", name="layer1"),
        layers.Dense(4, activation="relu", name="layer2"),
        layers.Dense(4, name="layer3"),
    ]
)
X = tf.ones((10, 10))
y = model(X)
```

You can see your model by using **model.layers**.

Code:

```
print(model.layers)
```

Output:

```
[<tensorflow.python.keras.layers.core.Dense object at 0x7f354f2609b0>,
<tensorflow.python.keras.layers.core.Dense object at 0x7f354c3b1278>,
<tensorflow.python.keras.layers.core.Dense object at 0x7f357bda4470>,
<tensorflow.python.keras.layers.core.Dense object at 0x7f354c2f28d0>]
```

Adding Layers in the Model

The **layers.add** method helps us add layers in the model. The argument to this function is a layer. For example, in the following code, a dense layer having 2 units and "relu" activation is added to the existing model. Note that **model.layers** outputs an extra layer.

Code:

```
model.add(layers.Dense(2, activation="relu"))
print(model.layers)
```

Output:

```
[<tensorflow.python.keras.layers.core.Dense object at 0x7f354f2609b0>,
<tensorflow.python.keras.layers.core.Dense object at 0x7f354c3b1278>,
<tensorflow.python.keras.layers.core.Dense object at 0x7f357bda4470>,
<tensorflow.python.keras.layers.core.Dense object at 0x7f354c2f28d0>,
<tensorflow.python.keras.layers.core.Dense object at 0x7f354c3b1668>]
```

Removing the Last Layer from the Model

The **layers.pop** method helps us pop a layer from the model. Since we intend to remove the last layer, we need not provide any argument to this function. For example, in the following code, the last layer is popped from the existing model.

Code:

```
model.pop()
```

Initializing Weights

The weights can be created only if the size of the input is known in advance. Initially, when the weights are not provided, there are no weights. The weights are created when the shape of the input is specified. The weights of a layer can be seen using the **layers.weights**. The following code creates a dense layer having ten neurons. When an input of size 5 × 5 is given to the layer, the shape of the weights so created becomes TensorShape([5, 10]). This is also applicable to sequential models.

Code:

```
model2=layers.Dense(10)
X=tf.ones((5,5))
y=model2(X)
model2.weights[0].shape
```

Output:

TensorShape([5, 10])

Summary

One can see the summary of a model using **model.summary()**. This method also displays the total number of parameters and the total number of learnable and non-learnable parameters.

Code:

```
model.summary()
```

Output:

Model: "sequential_4"

Layer (type)	Output Shape	Param #
layer1 (Dense)	(10, 5)	55
layer2 (Dense)	(10, 4)	24

```
layer3 (Dense)                    (10, 4)                           20
=================================================================
Total params: 99
Trainable params: 99
Non-trainable params: 0
```

Having seen the creation of a sequential model, let us now move to a brief discussion on Keras Layers.

Keras Layers

The Keras Layers application programming interface provides TITO (Tensor In Tensor Out) functions and the corresponding weights. In the training part, when a layer receives the data, the weights are stored in **layers.weights**. Some of the important layers of this interface are as follows.

You can import layers from the **tensorflow.keras**:

```
from tensorflow.keras import layers
```

1. Dense Layer

Name: layers.Dense

Function: This function creates a dense layer.

Most essential parameters: The number of output units and the activation

Example: In the following example, an output layer with ten neurons is created with the **relu** activation function. The shape of the input is (20, 20).

Code:

```
layer = layers.Dense(10, activation='relu')
inputs = tf.random.uniform(shape=(20, 20))
outputs = layer(inputs)
```

2. Conv2D Layer

Name: Conv2D

Function: The tf.keras.layers.Conv2D helps us create a Conv2D layer.

Parameters: The following syntax shows the parameters and their default values. Note that the strides parameter should be set to a tuple indicating strides. Likewise, the filters and kernel size can also be specified. Padding = "valid" indicates that the size of the output should be the same as that of the input.

Syntax:

```
tf.keras.layers.Conv2D(filters,kernel_size,strides=(2,2),padding="valid",activation=None,use_bias=True,bias_initializer="zeros")
```

3. Pooling

Name: MaxPooling2D

Function: This implements the max pooling operation for 2D spatial data.

Arguments: The **pool_size** must be set to the desired tuple indicating the size of the **pooling**. Here, strides can also be specified.

Syntax:

```
tf.keras.layers.MaxPooling2D(pool_size=(2,2),strides=None,padding="valid")
```

4. Activations

The activations, in the above layers, can be any of the following:

- relu function
- sigmoid function
- softmax function
- tanh function
- selu function
- exponential function

The syntax of the softmax and ReLU are as follows.

4.1 Softmax

Name: tf.keras.layers.Softmax
Function: This implements the softmax activation function.
Syntax:

```
tf.keras.layers.Softmax(axis=-1,**kwargs)
```

4.2 ReLU

Name: tf.keras.layers.ReLU
Function: This implements the ReLU activation.
Syntax:

```
tf.keras.layers.ReLU(max_value=None,negative_slope=0,threshold=0,**kwargs)
```

5. Initializing Weights

The weights can be initialized by any of the following classes:

- RandomNormal class
- RandomUniform class
- TruncatedNormal class
- Zeros class
- Ones class

Note that the initializations have also been dealt with in Chapter 3 of this book.

6. Miscellaneous

As in the case of Neural Networks, we can use L1 or L2 or L1-L2 regularizations. The constraints of the weights can also be specified using the layer's weight constraint class.

Having seen the building blocks of a sequential model and an overview of keras.layers, let us create a model to classify digits of the MNIST dataset.

CHAPTER 7 CONVOLUTIONAL NEURAL NETWORK: II

MNIST Dataset Classification Using LeNet: Prerequisite

Let us try to understand the layers by creating a simple dense network using ***keras.layers***. The reader is expected to write a code that implements a simple model to classify the MNIST dataset. The model should contain three layers having 20, 10, and 5 neurons, respectively. The Appendix A, given at the end of this book, discusses the training and evaluation of the models. The Appendix A also includes the code. However, try not to refer to the code before trying the task. You should get an accuracy of more than 90% using this model. Also, report the effect of change of the activation functions and the number of neurons in the hidden layer on the performance of the model.

The implementation of this task using LeNet is given in the next section. The reader is expected to compare the outputs of the two implementations, the number of trainable parameters, and the time required for training the models.

So far we have learned the creation of a model and its compilation. Let us now have a look at some of the most popular CNN models.

LeNet

LeNet was one of the first CNN models, which was successfully applied to handwritten digit recognition. This model laid the foundation of Convolutional Neural Networks. The model was developed at Bell Labs and applied the backpropagation algorithm to CNN. It had better generalization capabilities as compared with the single-layer networks as established by the paper [2] by the creator Yann LeCun. The proposed model displayed excellent performance, giving an error rate of just 1%.

Structure

The structure of this model would inspire many others and prove a milestone in Deep Learning research. Originally, the convolutions were referred to as the receptive fields. The pooling layers of this model perform average pooling. LeNet-5 had the following layers:

- The first layer of the model is a convolutional layer with six kernels of size 5 × 5.

- The next layer is a pooling layer, which converts the input to 14 × 14 by using a 2 × 2 average pooling.

- The next layer is a convolutional layer followed by a pooling layer similar to the second layer of this model.

- The fifth layer is a flattening layer, and the sixth layer is a fully connected layer with 120 units.

- The seventh layer is a fully connected layer with 84 units and an activation.

- The last layer is the output softmax layer, which is of size ten neurons.

Figure 7-1 shows the structure of LeNet.

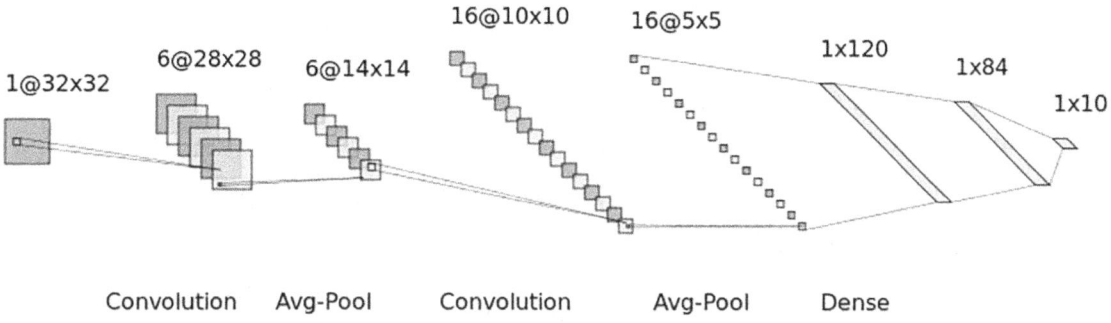

Figure 7-1. The structure of LeNet

Note that the size of the kernels in LeNet is small. This means the number of parameters to be learned is reduced, hence alleviating the performance of the network. This was necessary as these networks were designed in the late 1990s when the computational power of machines was limited. It may also be noted that the presence of small kernels hinders the capacity of the network to learn complex patterns, hence mitigating the chances of overfitting.

The choice of stride in a CNN also affects the performance of the model. The use of appropriate stride strikes a balance between the dimensionality of feature maps and the extraction of pertinent information. If the stride is too large, then it leads to loss of information.

CHAPTER 7 CONVOLUTIONAL NEURAL NETWORK: II

LeNet uses average pooling in place of max pooling, as max pooling extracts the most important part, whereas average pooling extracts the average information from the region, hence preserving the special structure.

Implementation

The following code implements the LeNet-5 model. The data is loaded and split into the train and the test set. Since the input images are 28 × 28 but LeNet takes 32 × 32 as input, padding is done. Note that the padding is done only in the second and the third dimensions as the first dimension represents the number of samples. The model is then created.

Step 1: This step involves loading the data and obtaining the train and the test data:

```
mnist_data=tf.keras.datasets.mnist
(X_train,y_train),(X_test,y_test)=mnist_data.load_data()
```

Step 2: Padding is done to convert 28 × 28 images to 32 × 32 images:

```
X_train=np.pad(X_train,((0,0),(2,2),(2,2)))
X_test=np.pad(X_test,((0,0),(2,2),(2,2)))
```

Step 3: In this step, the train and the test data are normalized, and the label is converted to a one-hot form:

```
X_train=np.reshape(X_train,(X_train.shape[0],32,32,1))
X_test=np.reshape(X_test,(X_test.shape[0],32,32,1))
X_train=X_train/255
X_test=X_test/255
X_train = X_train.astype('float32')
X_test = X_test.astype('float32')
y_train = tf.keras.utils.to_categorical(y_train, 10)
y_test = tf.keras.utils.to_categorical(y_test, 10)
```

Step 4: In this step, the model is crafted. This model consists of the following layers:

- The first layer is a convolutional layer with six filters having kernel size (5,5).
- The second layer is an average pooling layer of size (2,2).

- The third layer is a convolutional layer with 16 filters having kernel size (5,5).
- The fourth layer is an average pooling layer again of size (2,2).
- The fifth layer is a flattening layer.
- The sixth layer is a fully connected layer with 120 units and ReLU activation.
- The seventh layer is a fully connected layer with 84 units and ReLU activation.
- The last layer is the output softmax layer, which is of size ten neurons (one of ten digits).

```
model = tf.keras.Sequential()
model.add(tf.keras.layers.Conv2D(filters=6,kernel_size=(5, 5),strides=(1, 1),activation='tanh',input_shape=(32,32,1)))
model.add(tf.keras.layers.AveragePooling2D(pool_size=(2, 2),strides=(2, 2)))
model.add(tf.keras.layers.Conv2D(filters=16,kernel_size=(5, 5),strides=(1, 1),activation='tanh'))  model.add(tf.keras.layers.AveragePooling2D(pool_size=(2, 2),strides=(2, 2)))
model.add(tf.keras.layers.Flatten())
model.add(tf.keras.layers.Dense(units=120,activation='relu'))
model.add(tf.keras.layers.Dense(units=84, activation='relu'))
model.add(tf.keras.layers.Dense(units=10, activation='softmax'))
model.compile(loss='categorical_crossentropy',optimizer=SGD(lr=0.1),metrics=['accuracy'])
```

Step 5: This step involves training the model. Note that the batch size is taken as 128:

```
epochs = 10
history = model.fit(X_train, y_train,epochs=epochs,validation_data=(X_test,y_test),batch_size=128,verbose=2)
```

Output:

```
Epoch 1/10
469/469 - 29s - loss: 0.4425 - accuracy: 0.8658 - val_loss: 0.1771 - val_accuracy: 0.9427
Epoch 2/10
469/469 - 29s - loss: 0.1467 - accuracy: 0.9542 - val_loss: 0.1098 - val_accuracy: 0.9655
Epoch 3/10
469/469 - 29s - loss: 0.1043 - accuracy: 0.9687 - val_loss: 0.0930 - val_accuracy: 0.9703
Epoch 4/10
469/469 - 29s - loss: 0.0826 - accuracy: 0.9744 - val_loss: 0.0735 - val_accuracy: 0.9769
Epoch 5/10
469/469 - 29s - loss: 0.0686 - accuracy: 0.9783 - val_loss: 0.0603 - val_accuracy: 0.9799
Epoch 6/10
469/469 - 29s - loss: 0.0590 - accuracy: 0.9817 - val_loss: 0.0615 - val_accuracy: 0.9805
Epoch 7/10
469/469 - 29s - loss: 0.0512 - accuracy: 0.9843 - val_loss: 0.0727 - val_accuracy: 0.9769
Epoch 8/10
469/469 - 29s - loss: 0.0457 - accuracy: 0.9854 - val_loss: 0.0519 - val_accuracy: 0.9830
Epoch 9/10
469/469 - 29s - loss: 0.0414 - accuracy: 0.9870 - val_loss: 0.0457 - val_accuracy: 0.9844
Epoch 10/10
469/469 - 29s - loss: 0.0366 - accuracy: 0.9888 - val_loss: 0.0438 - val_accuracy: 0.9863
```

Step 6 (a): The training loss and the validation loss are then analyzed:

```
import matplotlib.pyplot as plt
num_epochs = np.arange(0, 10)
plt.figure()
plt.plot(num_epochs, history.history['loss'],label='Training Loss')
plt.plot(num_epochs, history.history['val_loss'],label='Validation Loss')
plt.xlabel('Epoch')
plt.ylabel('Loss')
plt.legend()
plt.show()
```

Output:

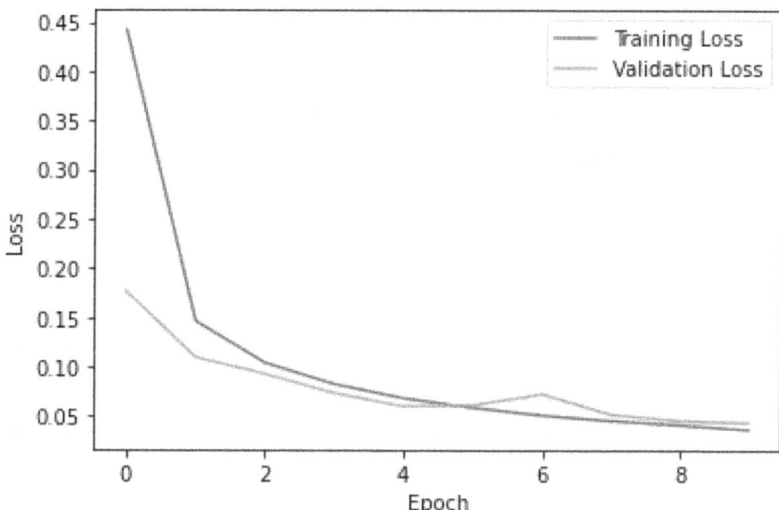

Step 6 (b): The training accuracy and the validation accuracy are then analyzed:

```
plt.figure()
plt.plot(num_epochs, history.history['accuracy'], label='Training
Accuracy')
plt.plot(num_epochs, history.history['val_accuracy'], label='Validation
Accuracy')
plt.xlabel('Epoch')
plt.ylabel('Accuracy')
plt.legend()
```

Output:

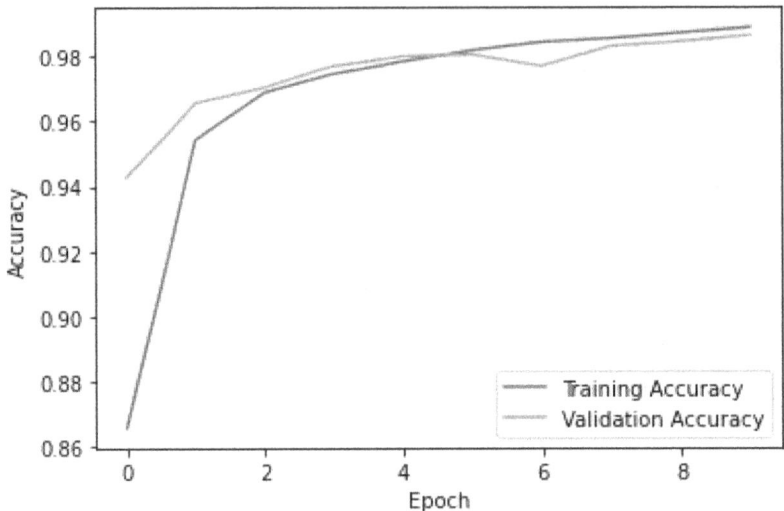

Having seen the architecture of LeNet and its application to the MNIST dataset, let us now move to another popular CNN, namely, AlexNet.

AlexNet

AlexNet was developed by Alex Krizhevsky, a Ukraine-born computer scientist. It is a CNN model, which won the ImageNet 2012 challenge. It was inspired by LeNet and had eight layers. This model used max pooling as against average pooling in LeNet-5. The model laid stress on the depth and used GPUs for training. AlexNet challenged the CNN models prevalent at that time, by reducing the training time, still improving the performance. The following features made AlexNet stand apart:

- Relu Activation: AlexNet used Rectified Linear Units instead of the popular sigmoid or the tanh function. This drastically reduced the time and helped this model achieve a 25% error rate on the CIFAR-10 dataset.

- GPU: AlexNet used multiple GPUs and divided the model neurons among them.

- Concept of Overlapping Pooling: The authors introduced the concept of overlapping pooling and established that this leads to less overfitting. This also improved its error rate.

CHAPTER 7   CONVOLUTIONAL NEURAL NETWORK: II

AlexNet was the winner of the ImageNet Large Scale Visual Recognition Challenge, in 2012. It had eight layers and was initially trained on more than 1,000,000 images. It could classify the images into 1000 classes and is generally considered better than LeNet. As per [3], the top 1% and top 5% error rates achieved with the help of this network were 37.5% and 17%, respectively. The total number of parameters in this network was around 60,000,000. The following discussion describes why this network performed much better than previously developed networks. The structure of this network is as follows.

Structure:

The model contains the Conv2D, activation, pooling, softmax, and dense layers, arranged as discussed in the code that follows.

Code:

```
model = Sequential()
model.add(Conv2D(filters=96, input_shape=(224,224,3), kernel_size=(11,11),
strides=(4,4), padding='valid'))
model.add(Activation('relu'))
model.add(MaxPooling2D(pool_size=(2,2), strides=(2,2), padding='valid'))
model.add(Conv2D(filters=256, kernel_size=(11,11), strides=(1,1),
padding='valid'))
model.add(Activation('relu'))
model.add(MaxPooling2D(pool_size=(2,2), strides=(2,2), padding='valid'))
model.add(Conv2D(filters=384, kernel_size=(3,3), strides=(1,1),
padding='valid'))
model.add(Activation('relu'))
model.add(Conv2D(filters=384, kernel_size=(3,3), strides=(1,1),
padding='valid'))
model.add(Activation('relu'))
model.add(Conv2D(filters=256, kernel_size=(3,3), strides=(1,1),
padding='valid'))
model.add(Activation('relu'))
model.add(MaxPooling2D(pool_size=(2,2), strides=(2,2), padding='valid'))
model.add(Flatten())
model.add(Dense(4096, input_shape=(224*224*3,)))
model.add(Activation('relu'))
model.add(Dropout(0.4))
model.add(Dense(4096))
```

```
model.add(Activation('relu'))
# Add Dropout
model.add(Dropout(0.4))
model.add(Dense(1000))
model.add(Activation('relu'))
model.add(Dropout(0.4))
model.add(Dense(10))
model.add(Activation('softmax'))
```

The major problem with the above architecture is the inability of a single GPU (GTX 580) available at that time to house all the training data, as they needed 1.2 million examples to train the network. To handle this they used multiple GPUs (two in particular) and divided the kernels among them. For example, in the first convolutional layer, the number of kernels was 96, and each GPU was provided with 48 of them. Likewise, for the second each was provided with 128 of them. The third, fourth, and fifth were also divided accordingly. Refer to Figure 7-2 [1] in which the parallelization scheme is shown.

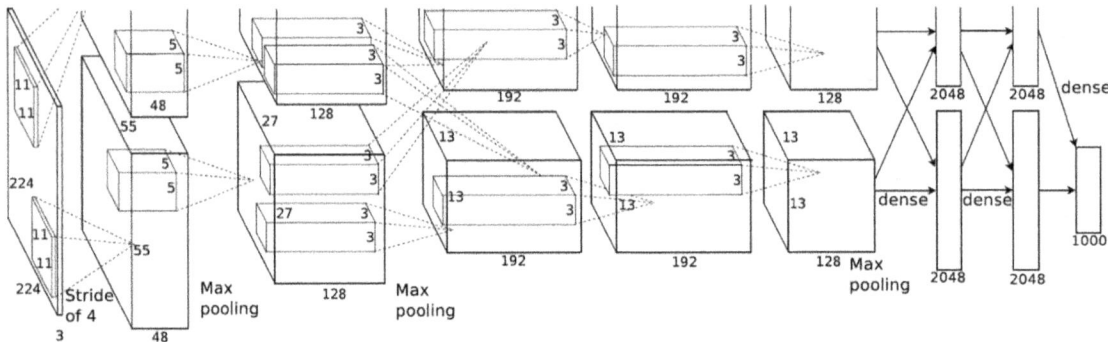

Figure 7-2. *The structure of AlexNet as shown in the original paper [1]*

As per the paper, the above trick reduced the top 1% error by 1.7%. The author also employed local response normalization in which the activity of a neuron was normalized vis-a-vis the adjacent kernels "at the same spatial positions." They also employed overlapping pooling as against the conventional non-overlapping pooling used in the earlier architectures.

In order to reduce overfitting, two major tricks were used: one was data augmentation, which effectively means to artificially increase the dataset by preserving the corresponding labels. They use the concept of generating the augmented data and not storing in the memory. The other method they employed for reducing the overfitting was dropout. They used p = 0.5 in each of the hidden neurons. This reduced the overfitting but doubled the number of iterations required for coverage.

This architecture was far better than LeNet, but more effective and efficient architectures were yet to come. The next section presents some such architectures.

Some More Architectures
GoogLeNet

GoogLeNet was the winner of the 2014 ILSVRC competition. The model was originally called Inception V1, as it introduced the inception block. The block uses three filters of sizes ranging from 1×1 to 5×5. This allowed the model to capture the course as well as the finer details. It may be noted that the model confirms the computations by adding a bottleneck of 1×1. The model also used sparse connections and normalization. The global average pooling, in the last layer, and the RmsProp optimizer were used in the model. The model was heterogeneous, and the topology needed management in each module. The feature space was being drastically reduced in the next layer, therefore leading to the possibility of a loss of important information.

The inception module includes the combination of convolutional layers having different kernel sizes and pooling layers. These multiple branches extract the features at different scales, thus generating a richer set of features, hence recognizing the complex patterns. This also makes the network more efficient.

ResNet

ResNet was the winner of the 2015 ILSVRC competition. It is a 152-layer-deep CNN. Despite being deeper as compared with AlexNet and VGG, it demonstrated lesser computational complexity. This model introduced the concept of residual learning. As a matter of fact, ResNet gained a 28% improvement on the famous image recognition benchmark dataset named COCO. The idea of bypassing the pathways used in Highway Networks was exploited in the model to address the issues in training the networks. ResNet introduced shortcut connections within layers to enable cross-layer connectivity. This sped up the convergence of deep networks, thereby providing the ability to avoid gradient diminishing problems to the ResNet.

DenseNet

This model was conceived to solve the vanishing gradient problem. This model connected each preceding layer to all the next layers in a feed-forward fashion. This implies that the feature maps of all the previous layers were used as inputs into all subsequent layers [4]. This provides the ability to explicitly differentiate between information that is added to the network and information that is preserved to the network. However, this model is parametrically pricey especially on increasing the number of feature maps. Take a pause and think if this CNN can be considered a sequential model.

Conclusion

CNNs are generally used for images. Their performance on images is far superior vis-à-vis the MLPs. The last two chapters discussed the basics of CNNs, various models, and implementations. The chapters presented implementations from scratch and the use of Keras. The reader should be able to implement the models and make changes in the existing models. However, in some of the models, the number of learnable parameters is huge. In the next chapters, we will learn how to deal with this problem.

Having learned MLPs and CNNs, the next chapters deal with the applications of these CNNs in object recognition and segmentation. Before that let's test what we have learned.

Exercises

Multiple-Choice Questions

1. In which of the following cases sequence models cannot be used?

    a. If the model has multiple inputs, then the sequential model is not used.

    b. In the case of a model having multiple outputs.

    c. In the case of nonlinear models.

    d. All of the above.

CHAPTER 7 CONVOLUTIONAL NEURAL NETWORK: II

2. Which method helps us pop a layer from the model?

    a. model.pop()

    b. model.add()

    c. model.summary()

    d. None of the above

3. The Keras Layers application programming interface provides

    a. TITO (Tensor In Tensor Out) functions

    b. LIFO

    c. FIFO

    d. None of the above

4. Which of the following activations can be used in sequential models?

    a. relu function

    b. sigmoid function

    c. softmax function

    d. tanh function

    e. All of the above

5. The weights can be initialized by which of the following classes?

    a. RandomNormal class

    b. RandomUniform class

    c. TruncatedNormal class

    d. All of the above

6. Which model is generally considered as one of the successful models that was applied to handwritten digit recognition?

    a. LeNet

    b. AlexNet

    c. Google LeNet

    d. None of the above

7. Which model first used the combination of ReLU, GPU power, and overlapping pooling?

    a. AlexNet

    b. LeNet

    c. Google LeNet

    d. None of the above

8. Which model introduced the inception block?

    a. Google LeNet

    b. AlexNet

    c. LeNet

    d. None of the above

9. Which model introduced the concept of residual learning?

    a. DenseNet

    b. ResNet

    c. AlexNet

    d. None of the above

10. Which of the following was conceived to solve the vanishing gradient problem?

    a. DenseNet

    b. AlexNet

    c. LeNet

    d. None of the above

Implementations

Refer to the following datasets. Use the models of Question 1 and compare the performance of the models. Also, reduce or increase the depth and report the effect on the performance. State some of the measures that you would take to handle the bias and the variance in the so-developed model and report your results. Figure out why some of the stated methods worked with the given datasets:

1. https://www.kaggle.com/puneet6060/intel-image-classification

2. https://www.kaggle.com/vishalsubbiah/pokemon-images-and-types

3. https://www.kaggle.com/shravankumar9892/image-colorization

4. https://www.kaggle.com/hsankesara/flickr-image-dataset

References

[1] Krizhevsky, Alex; Sutskever, Ilya; Hinton, Geoffrey E. (2017-05-24). "ImageNet classification with deep convolutional neural networks" (PDF). Communications of the ACM. 60 (6): 84–90. doi:10.1145/3065386. ISSN 0001-0782. S2CID 195908774.

[2] Lecun, Y.; Bottou, L.; Bengio, Y.; Haffner, P. (1998). "Gradient-based learning applied to document recognition" (PDF). Proceedings of the IEEE. 86 (11): 2278–2324.

[3] Papers with Code - AlexNet Explained. https://paperswithcode.com/method/alexnet

[4] Huang, G., Liu, Z., Laurens, V. D. M., & Weinberger, K. Q. (2016, August 25). Densely connected convolutional networks. arXiv.org. https://arxiv.org/abs/1608.06993v5

CHAPTER 8

Transfer Learning

Introduction

Assume that you have been assigned the responsibility of developing an app that can classify ten new musical instruments developed recently by renowned musicians. You only have a few hundred images of these instruments. To classify these images, you decide to use GoogLeNet architecture, which has around 6.8 million parameters. If you decide to train the model using the given images, you will realize that you have an insufficient number of images. However, if you train the model using the pictures of known instruments, you might be able to train it, but it will take a lot of time and computational resources. So the challenge is to train a sufficiently complex model on a dataset that does not have a sufficient number of images, and you probably do not have GPU as well.

This chapter presents a methodology called transfer learning, which will help you deal with such situations. This chapter discusses the ideas, types, strategies, and limitations of transfer learning.

Idea

In transfer learning, we train a model on a given dataset for a particular task. This model is then

   i) Used with some other dataset for the same task. For example, suppose you aim to develop a model that classifies patients suffering from Alzheimer's from controls. You can train the model on the publicly available Alzheimer's Disease Neuroimaging Initiative (ADNI) dataset and then use the same model on the dataset collected from a local hospital.

CHAPTER 8   TRANSFER LEARNING

ii) We use the same model and the same dataset for some other task. For example, we develop a model to classify cats and dogs using given images and then use the same model by some means for segmenting the parts of the image. The interested readers may refer to references given at the end of the chapter for such examples of this type of transfer learning.

iii) We use the part of the same model and a new dataset for different tasks. For example, we generally use the pretrained VGG 16 model and freeze the initial layers except for the last ones (fully connected), and we train the model to perform the classification on other datasets.

One of the ways to accomplish the task stated in "Introduction" is "**to extract knowledge from some model trained on some dataset and use it to accomplish a similar task**." This is referred to as transfer learning.

This is possible because the earlier layers of a complex model trained on sufficiently large datasets learn **low-level features**, the next layers may learn the **combination of such features,** and so on. To understand this, imagine you develop a model and train it using a huge dataset of faces. The earlier layers of the model learn lines, curves, etc. The later layers learn objects; still later layers learn eyes, nose, etc. You can use this information to classify some other dataset of faces for a particular organization to develop their face recognition system.

VGG 16 and VGG 19 for Binary Classification

VGG 16 and VGG 19 are two deep Convolutional Neural Networks having 16 and 19 layers, respectively (trainable). They have historically outperformed the benchmarks for many image-related tasks. The VGG 16 model is an outcome of the work "Very Deep Convolutional Networks for Large-Scale Image Recognition" [1].

VGG 16 achieved 92.7% top 5 test accuracy on the ImageNet dataset containing 14 million images belonging to 1000 classes. This model takes 224 × 224 × 3 as the input. It contains two convolutional layers with a filter size of 3 × 3 followed by a max pooling layer with a filter size of 2 × 2. This is repeated twice. After which it contains three convolutional layers with a filter size of 3 × 3 followed by a single max pooling layer of filter size 2 × 2, and this combination is repeated thrice. This is followed by two fully connected layers of size 4096 followed by an output layer having 1000 neurons. VGG 19 has a similar architecture, but it has 19 layers instead of 16. Figures 8-1 and 8-2 show the architecture of VGG 16 and VGG 19.

CHAPTER 8 TRANSFER LEARNING

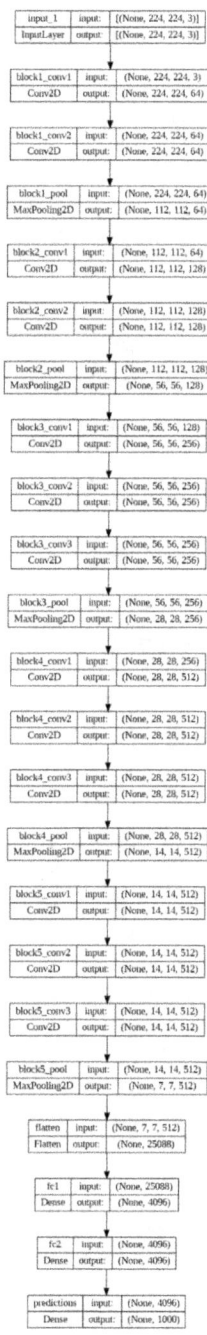

Figure 8-1. *VGG 16 architecture*

CHAPTER 8　TRANSFER LEARNING

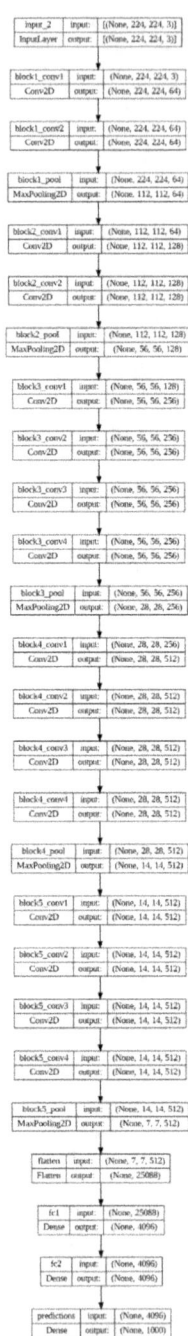

Figure 8-2. *VGG 19 architecture*

CHAPTER 8   TRANSFER LEARNING

VGG 16 and VGG 19 have been trained on huge datasets (ImageNet) having a massive number of images across a thousand categories. This training gives the model the ability to learn generalizable features and empowers these models with the ability to capture complex patterns; hence, they perform well with other datasets as well.

To be able to use the information gathered in the training process, we generally freeze the earlier layers and train the last layers of these models. These models have shown good performance on many image-related tasks.

In the following experiment (Listing 8-1), the pretrained VGG 16 and VGG 19 models are to classify X-ray images of patients diagnosed with tuberculosis (TB) and healthy controls. The dataset, obtained from Kaggle ("https://www.kaggle.com/datasets/tawsifurrahman/tuberculosis-tb-chest-xray-dataset"), contains 400 images of healthy controls and 240 images of TB patients.

The given images were resized to 224 × 224 × 3 shape to match the input shape of the original model. The initial layers of the pretrained models were frozen to extract low-level features. The last few layers were then trained on the above mentioned dataset to learn high-level, data-specific features that distinguish between TB patients and controls. The loss and performance curves of both the models are shown in Figures 8-3 and 8-4.

Listing 8-1. Binary classification using VGG 16 and VGG 19

```
#1. Import the required libraries
import numpy as np
import pandas as pd
from matplotlib import pyplot as plt
import tensorflow as tf
from tensorflow.keras.applications import VGG16, VGG19
from tensorflow.keras.models import Model
from tensorflow.keras.layers import Dense, Dropout, Flatten
from sklearn.model_selection import train_test_split
#2. Load the dataset
X = np.load('/content /X.npy')
y = np.load('/content /y.npy')
#3. Split the dataset into train and test set
X_train, X_test, y_train, y_test = train_test_split(X, y, test_size=0.3)
```

CHAPTER 8   TRANSFER LEARNING

#4. Load the pre-trained models
```
base_model_vgg16 = VGG16(weights='imagenet', include_top=False, input_shape=(224, 224, 3))
base_model_vgg19 = VGG19(weights='imagenet', include_top=False, input_shape=(224, 224, 3))
```
#5. Freeze the initial layers
```
for layer in base_model_vgg16.layers:
    layer.trainable = False
for layer in base_model_vgg19.layers:
    layer.trainable = False
```
#6. Create a function to add dense layers for binary classification
```
def add_custom_layers(base_model):
    x = base_model.output
    x = Flatten()(x)
    x = Dense(1024, activation='relu')(x)
    x = Dropout(0.5)(x)
    predictions = Dense(1, activation='sigmoid')(x)   # Example for 10 classes
    return Model(inputs=base_model.input, outputs=predictions)
```
#7. Initialize the new models
```
model_vgg16 = add_custom_layers(base_model_vgg16)
model_vgg19 = add_custom_layers(base_model_vgg19)
```
#8. Compile and fit the above models
```
model_vgg16.compile(optimizer='adam', loss='binary_crossentropy', metrics=['accuracy'])
model_vgg19.compile(optimizer='adam', loss='binary_crossentropy', metrics=['accuracy'])
history_1 = model_vgg16.fit(X_train, y_train, epochs=10, batch_size=32, validation_data=(X_test, y_test))
history_2 = model_vgg19.fit(X_train, y_train, epochs=10, batch_size=32, validation_data=(X_test, y_test))
```
#9. Create a function to plot loss and accuracy curve
```
def plot_history(history, model_name):
    plt.figure(figsize=(12, 6))
    plt.subplot(1, 2, 1)
    plt.plot(history.history['accuracy'])
```

```python
    plt.plot(history.history['val_accuracy'])
    plt.title(f'{model_name} Model Accuracy')
    plt.xlabel('Epoch')
    plt.ylabel('Accuracy')
    plt.legend(['Train', 'Val'], loc='upper left')
    plt.subplot(1, 2, 2)
    plt.plot(history.history['loss'])
    plt.plot(history.history['val_loss'])
    plt.title(f'{model_name} Model Loss')
    plt.xlabel('Epoch')
    plt.ylabel('Loss')
    plt.legend(['Train', 'Val'], loc='upper left')
    plt.tight_layout()
    plt.show()
#10. Plotting accuracy and loss curves for each model
plot_history(history_1, "Model VGG16")
plot_history(history_2, "Model VGG19")
```

Output:

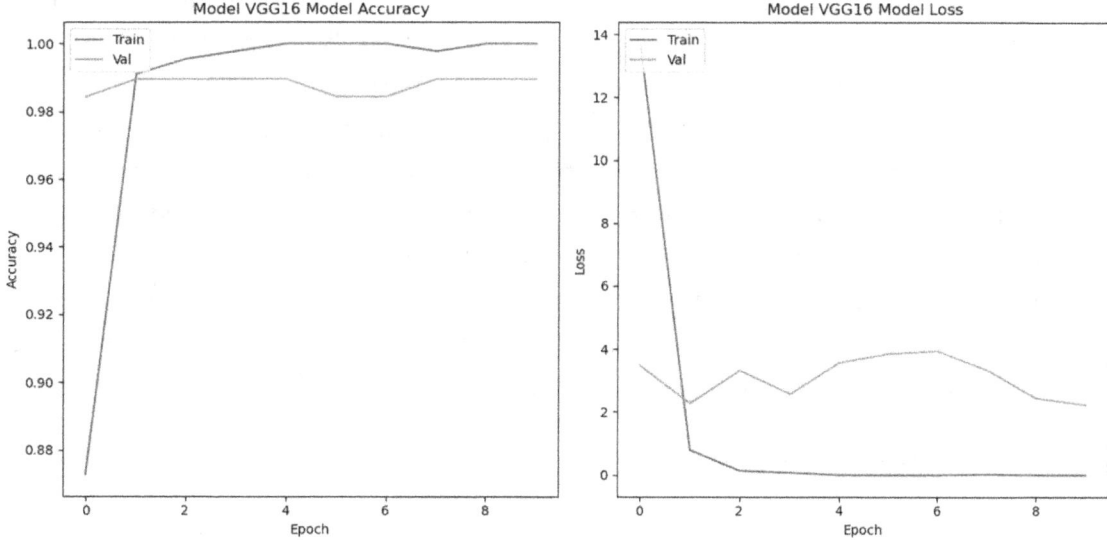

Figure 8-3. Loss and accuracy curves: VGG 16

CHAPTER 8 TRANSFER LEARNING

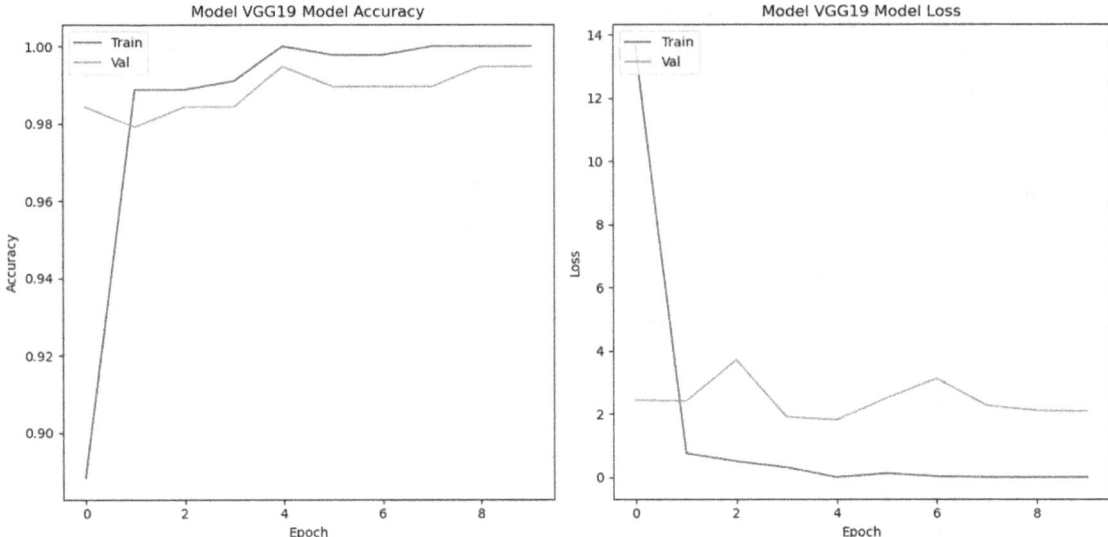

Figure 8-4. Loss and accuracy curves: VGG 19

From the above figures, it can be observed that the mean validation accuracy of the two models is 0.9880 (VGG 16) and 0.9886 (VGG 19), respectively. Also, there is a slight difference between the loss curves of the two models.

Let us take another example to understand the applications of transfer learning. The following experiment (Listing 8-2) employs a transfer learning approach to classify Alzheimer's patients from controls using the OASIS-1 dataset. The dataset includes s-MRI scans of 53 controls and 28 patients suffering from Alzheimer's disease (AD). The grayscale images were resized to (224 × 224). The additional Conv2D layer is added to convert the single-channel input to three channels by repeating the grayscale information across three channels to match the input shape of the pretrained VGG 19 model. The initial layers of the pretrained model were frozen to extract low-level features. The last few layers were then trained on the abovementioned dataset to learn high-level, data-specific features that distinguish between AD patients and controls. The loss and performance curves of the model are shown in Figure 8-5.

Listing 8-2. Alzheimer's classification using VGG 19

Code:
#1. Import the required libraries
```
import tensorflow as tf
from tensorflow.keras.applications import VGG19
from tensorflow.keras.models import Model
```

```python
from tensorflow.keras.layers import Input, Conv2D
from tensorflow.keras.layers import Flatten, Dense, Dropout
from tensorflow.keras.optimizers import Adam
import numpy as np
from sklearn.model_selection import train_test_split
import tensorflow as tf
from tensorflow.keras import datasets, layers, models
import matplotlib.pyplot as plt
```
#2. Load the dataset
```python
X = np.load('/content /X.npy')
y = np.load('/content /y.npy')
```
#3. Split the dataset into train and test set
```python
X_train, X_test, y_train, y_test = train_test_split(X, y, test_size = 0.3, shuffle = True)
print(X_train.shape, y_train.shape, X_test.shape, y_test.shape)
```
#4. Load the pre-trained models and freeze the initial layers
```python
base_model = VGG19(weights='imagenet', include_top=False, input_shape=(224, 224, 3))
for layer in base_model.layers:
    layer.trainable = False
```
#5. Create a new input layer for grayscale images
```python
new_input = Input(shape=(224, 224, 1))
```
#6. Add a Conv2D layer to convert grayscale images to 3 channels
```python
x = Conv2D(3, (3, 3), padding='same')(new_input)
x = base_model(x)
x = Flatten()(x)
x = Dense(1024, activation='relu')(x)
x = Dropout(0.5)(x)
x = Dense(1, activation='sigmoid')(x)
```
#7. Create, compile and fit the new model
```python
model = Model(inputs=new_input, outputs=x)
model.compile(optimizer=Adam(),loss='binary_crossentropy', metrics=['accuracy'])
model.summary()
batch_size = 64
```

CHAPTER 8 TRANSFER LEARNING

```
history_batch = model.fit(X_train, y_train, epochs=10, batch_size=batch_size, validation_data=(X_test, y_test))
```
#8. Create a function to plot loss and accuracy curve
```
def plot_history(history, model_name):
    plt.figure(figsize=(12, 6))
    plt.subplot(1, 2, 1)
    plt.plot(history.history['accuracy'])
    plt.plot(history.history['val_accuracy'])
    plt.title(f'{model_name} Model Accuracy')
    plt.xlabel('Epoch')
    plt.ylabel('Accuracy')
    plt.legend(['Train', 'Val'], loc='upper left')
    plt.subplot(1, 2, 2)
    plt.plot(history.history['loss'])
    plt.plot(history.history['val_loss'])
    plt.title(f'{model_name} Model Loss')
    plt.xlabel('Epoch')
    plt.ylabel('Loss')
    plt.legend(['Train', 'Val'], loc='upper left')
    plt.tight_layout()
    plt.show()
```
#9. Plot accuracy and loss curve for the above model
```
plot_history(history_batch, "Model VGG19")
```

Output:

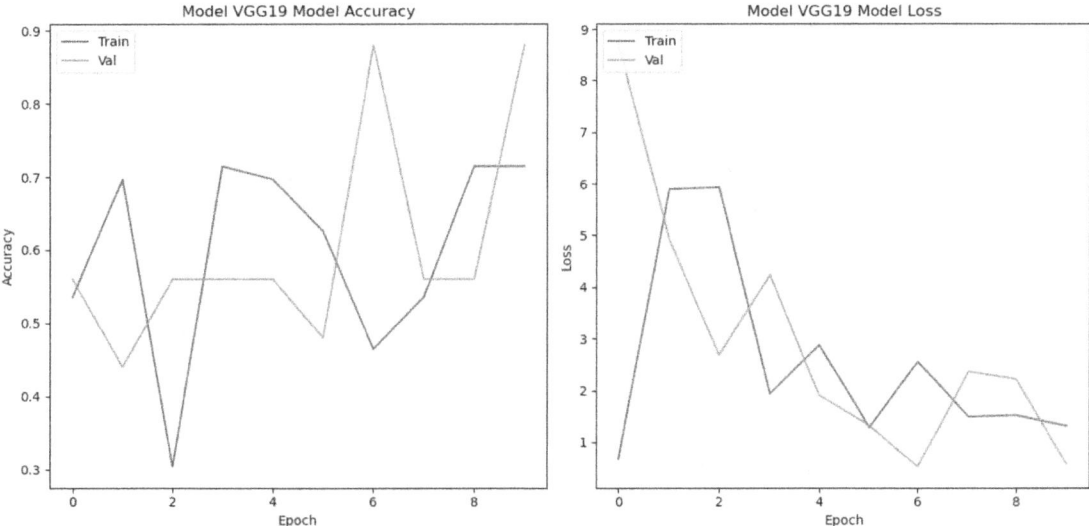

Figure 8-5. Loss and accuracy curves: VGG 19

Types and Strategies

An important aspect of transfer learning is its ability to transform the representation. One of the interesting examples as stated in [2] is as follows.

Assume that you need to classify two classes in which the 2D coordinate system is represented as a circle (Class I) within another circle (Class II) and somehow this model transforms this distribution to a linearly separable one; then the classification will become easier. At times transfer learning transforms the given data space into a feature set relatively easy to classify.

As per [2], the domain of transfer learning consists of a feature space and a probability distribution, whereas the task contains the label space. The probability $P(X|Y)$ is derived from a function that learns from a feature vector and the label space. Transfer learning can also be segregated into the following types:

1. When the feature spaces are not equal: Suppose that you are asked to develop a software that converts a given algorithm into a code in a functional language such as Scala or F#. Note that the input to the software is in English, whereas the output is in complex functional language. In this case the feature spaces are

not the same. Such situations can be handled using a concept similar to cross-lingual adaptation. Here transfer learning comes to your rescue.

2. When the marginal probability distributions of the source and the target are not same: Consider a scenario in which you are required to develop a model to distinguish between cats and dogs. The model is then trained on high-definition pictures obtained from the Internet. The application so developed is intended to be used by people in the lower middle strata who click the pictures of dogs and cats using their phone, which are not of high quality. In this situation the probability distributions of the source and the target are not same, and transfer learning can help.

3. When the labels are different: Suppose you train your model on images of animals and the model is to be used for classifying different types of cats. Here, the labels of the model originally developed with the original dataset have different labels vis-à-vis the required model.

4. When the conditional probability of the labels is not the same: When you train your model with a balanced dataset and then use it for an imbalanced one.

Transfer learning is tricky and whether to use it or not can be decided based on

1. The task to be performed
2. Domain
3. Availability of data

For a detailed discussion on the above strategies, the interested readers may refer to the references given at the end of the chapter. Researchers [2] have proposed many transfer learning strategies as shown in Figure 8-6.

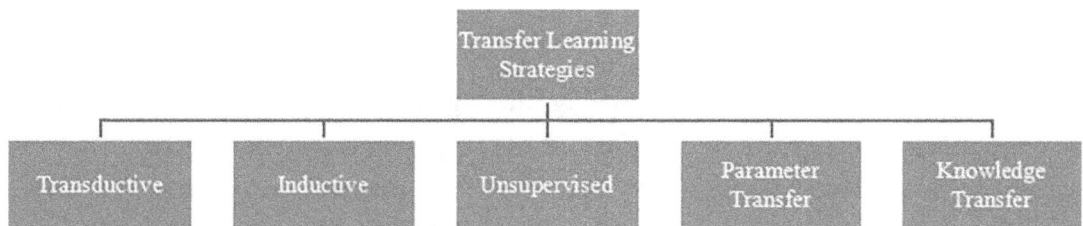

Figure 8-6. *Transfer learning strategies*

The interested reader may refer to the references given at the end of the chapter.

Limitations and Applications of Transfer Learning

Despite being awesome transfer learning has many limitations. If the target dataset does not have anything similar to the original dataset, then the transfer learning will not work. For example, if you train your model with the images of dogs and cats and then test it on a particular disease related to the brain, then the model is bound to fail. Likewise, if the number of labels in your target set is huge, then the model might not work as well. Some of the cases where transfer learning may not work are as follows:

- When the training data is insufficient, transfer learning may not work. In some cases, the training data might not be similar to the data used for the task at hand, or there is a domain mismatch or task mismatch. In such cases, the transfer learning generally fails.

- In addition to the above, the size of the target data also decides whether or not we can use transfer learning. If the size of the target data is small, there is a possibility of overfitting; also if the target data is too large, transfer learning may not be able to capture the complexities of the data. In transfer learning, freezing of incorrect layers may also affect the recital of the network.

Some of the prominent applications of transfer learning are

- Classification of diseases using models trained on similar diseases
- Task related to self-driven cars
- Natural language processing
- Identifying rare elements and so on

Conclusion

It is commonly believed that Deep Learning can only be applied if there is a lot of data. Also, as per common perception, a lot of computing power is consumed to train a DL [3, 4] model. However, for many practical tasks, this may not be required. We can learn micro-level or intermediate-level features from a particular source and apply the knowledge so obtained on another task or another dataset [5]. This chapter introduced transfer learning and explained the need, types, and implementation of transfer learning. After reading this chapter, the reader must have realized the need for huge datasets for carrying out tasks that are not related to the ones for which those datasets were collected. We have seen how to extract the representations and how to fine-tune a model for carrying out some of the assorted tasks using transfer learning.

The next chapter takes the reader to the mesmerizing world of sequences, where words play with each other and spin prose and poetry. We will study the models that will help us comprehend sequences and play with them.

Exercises

Multiple-Choice Questions

1. Which of the following exemplifies the scenario when feature spaces are not equal in transfer learning?

    a. Translating text from English to Spanish

    b. Converting an algorithm into code in a functional language such as Scala or F#

    c. Translating a book from English to French

    d. Converting a mathematical equation into a graph

2. What scenario exemplifies when the marginal probability distributions of the source and target are not the same?

    a. Translating an algorithm to code in a functional language

    b. Developing a model to distinguish between cats and dogs with high-definition pictures and using it on low-quality phone pictures

c. Classifying different types of cats using a model trained on images of animals

d. Translating a book from one language to another

3. What factor does NOT affect the decision to use transfer learning?

   a. The task to be performed

   b. Domain

   c. Availability of data

   d. The programming language used

4. Which statement about transfer learning is TRUE?

   a. It always requires a large amount of data.

   b. It can help when the feature spaces are equal.

   c. It makes it possible to develop a Deep Learning model with less data and computation power.

   d. It is only useful in natural language processing tasks.

5. What are the benefits of transfer learning mentioned in the text?

   a. It requires more data and computation power.

   b. It allows for classification with less data and less computation power.

   c. It eliminates the need for training models.

   d. It always produces higher-accuracy models.

6. Which type of transfer learning involves developing a software that converts a given algorithm into a code in a functional language?

   a. When feature spaces are not equal

   b. When marginal probability distributions are not the same

   c. When labels are different

   d. When conditional probability of the labels is not the same

CHAPTER 8   TRANSFER LEARNING

Application

1. Collect 100 images each of the following characters of the popular show *Phineas and Ferb*:

    a. Phineas Flynn

    b. Ferb Fletcher

    c. Candace Flynn

    d. Perry the Platypus

    e. Dr. Heinz Doofenshmirtz

    f. Isabella Garcia-Shapiro

    g. Baljeet Tjinder

    h. Buford van Stomm

    i. Linda Flynn-Fletcher

    j. Lawrence Fletcher

    k. Major Monogram

    l. Carl the Intern

2. Now, create a CNN model to classify the above classes. Now use pretrained VGG 16 and VGG 19 models to classify the images. You may use a different number of neurons in the fully connected layers and report the performance.

References

[1] Simonyan, K. & Zisserman, A. Very deep convolutional networks for Large-Scale image recognition. *arXiv (Cornell University)* (2014). https://doi.org/10.48550/arxiv.1409.1556

[2] Protopapas, P. Intro to transfer learning. In *Advanced Practical Data Science* (p. AC295) (2021). https://harvard-iacs.github.io/2020F-AC295/lectures/lecture5/presentation/lecture5.pdf

[3] Johnson, J. (2020, October 7). *Lecture 11: Training Neural Networks (Part 2)*. https://web.eecs.umich.edu/~justincj/slides/eecs498/FA2020/598_FA2020_lecture11.pdf

[4] Pesah, A., Wehenkel, A., Sedghi, H., Liang, P. & Zeng, D. *CS 330 Lecture 3 Transfer Learning + Start of Meta-Learning*. https://web.stanford.edu/class/cs330/lecture_slides/cs330_transfer_meta_learning.pdf

[5] Layton, O. *CS 343 | Notes*. https://cs.colby.edu/courses/F22/cs343/notes.html

CHAPTER 9

Recurrent Neural Network

Introduction

Consider a person named Nishant, about to go to meet someone important. Just by looking at the image in Figure 9-1, can you guess where will he head? Your answer would be a random guess at best. Now if you are given the position of this person in the past five time stamps and you need to guess the position at the next time stamp, your work will become a little easier. Your answer is now based on a sequence depicting the position at various time intervals.

Figure 9-1. *If no context is given, can you guess where will the person move next? (Image generated by* `https://pixlr.com/image-generator/`*)*

CHAPTER 9 RECURRENT NEURAL NETWORK

In many cases, if the values of a sequence at previous time stamps are known, then it becomes easy to guess the position at the next time stamp. Can you apply the same analogy to the prices of a stock, given its prices in the previous few weeks? It turns out that this is practically possible. Likewise, elements of music and text data also constitute sequences. To handle sequence data, we need slightly different kinds of models. To appreciate the need of a different kind of model, let us first try to handle this problem using Neural Networks.

Assume that $X_1, X_2...X_k$ are the values of the sequence at various time stamps. To predict the value of the sequence at the next time stamp, we create a Neural Network shown in Figure 9-2. You train this network using such sequences. However, the network might not perform well. (Why?)

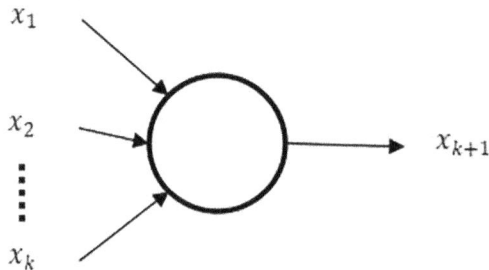

Figure 9-2. *Handling sequence data using Neural Networks*

To accomplish the above task, we need special types of networks constituting independent units in which x_0 predicts y_0, x_1 predicts y_1, and x_2 predicts y_2. (At time stamp 0 the value of the input is x_0 and the value of the output is y_0. Likewise, at time stamp 1 the value of the input is x_1 and the value of the output is y_1.) If we have to predict y_k, then it does not only depend on x_k; it may also depend on the previous inputs. Let us go a little deep!

Why Neural Networks Cannot Infer Sequences

Consider a sequence $\{x_1, x_2, x_3, ...x_n\}$, where x_1 is the value at time t_1, x_2 is the value at time t_2, and so on. We aim to design a model that understands this sequence. That is, predict values at the next time stamp. We start with a fully connected neural network, that takes k values and predicts the next value. For example, if the value of k is 4, then the model takes $\{x_1, x_2, x_3, x_4\}$ as input and predicts x_5; then takes $\{x_2, x_3, x_4, x_5\}$ as input and predicts x_6; and so on (Figure 9-3). To accomplish this task, we create a Neural Network having four neurons in the input layer and a single neuron at the output layer.

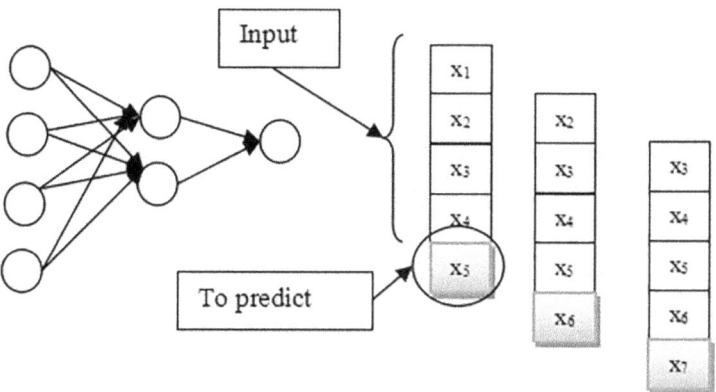

Figure 9-3. *Neural Network to predict the next element of the sequence. At t=1, the first four elements of the sequence are given as the input, and the network predicts the fifth element.*

The division of input data in such a manner is referred to as the **overlapping window**. The model, if trained with a sufficient amount of data, may start predicting the next element. However, if the order of the elements changes (say $\{x_3, x_2, x_1, x_4\}$), the model still predicts the same value (x_5). The reason for this is that the Neural Networks do not understand the context. However, for a programmer handling sequence data, this can be disastrous. For example, consider the following part of the sentence and try predicting the next word:

"In a place called Shangri-La, a person kills the son of one of the richest persons by his speeding car. He should go to …"

Here "jail" should be the next word that is obvious. However, for the following sentence

"In a place called Shangri-La, the son of one of the richest people kills a person by his speeding car. He should go to …"

since this is Shangri-La, the next word is not obvious; it can be "jail" or "essay-writing-classes." So a fully connected network may not be able to generate the correct (expected) answer.

This is because, for a fully connected neural network, the output is some function of inputs. The sequence models discussed in this chapter can infer the patterns in a sequence and extract temporal information from it. As stated earlier, sequences are

CHAPTER 9 RECURRENT NEURAL NETWORK

found everywhere, from text to sound and to time series. In addition to the above, there is another prominent difference between Neural Networks and sequence models, which is that the sequence models can handle the variable-length dataset.

Idea

A unit in a sequence model is expected to extract the context of a particular element, so it should remember some information regarding the earlier elements of the sequence. That is, it should have memory. We can use a recurrent unit to accomplish this task. In the recurrent unit, we give the input, it produces an output, and there is a hidden state. Let the weight associated with the input be W_{xh}, that with the output be W_{yh}, and that with the hidden state be W_{hh}. With each input, these weights are updated.

The unit of a Recurrent Neural Network (RNN) (Figure 9-4) can be considered as having an input $x^{<t>}$ associated with the weight W_{xh}, an output $y^{<t>}$ associated with the weight W_{hy}, and the hidden state $h^{<t>}$ associated with the weight W_{hh}. Note that the input and output change with the time "t," whereas the weights remain the same at all the time stamps. When we unwind it for different time stamps, we get an architecture as shown in Figure 9-4.

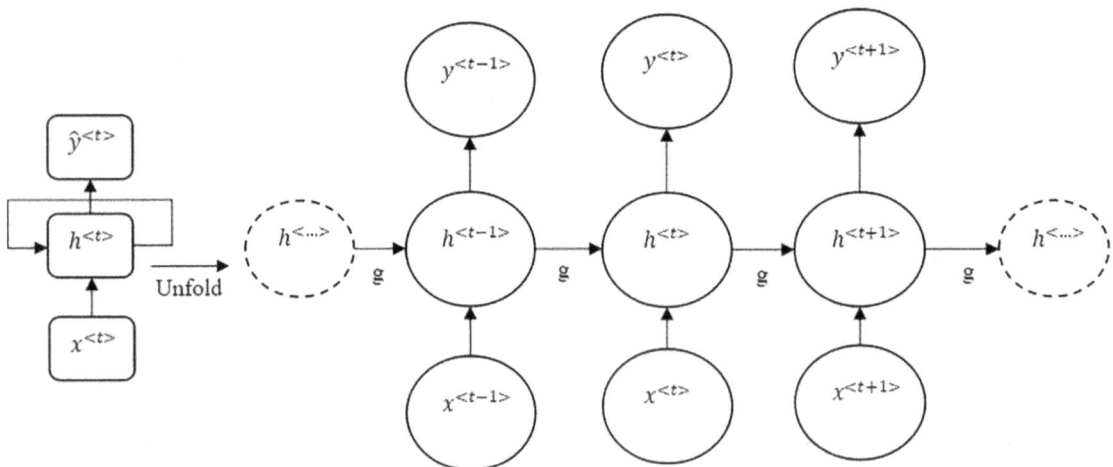

Figure 9-4. *A Recurrent Neural Network*

Note that the above is just one type of architecture; there are three more types of RNN, discussed in the following sections. The weights in this architecture are updated using an algorithm called Backpropagation Through Time (BPTT), discussed in the next section.

Backpropagation Through Time

Consider a sequence of length "T" and apply the forward and backward pass only on the unfolded network (K units). As stated earlier, we have two activation functions, one for the hidden state and one for the output. The value of the hidden state at time t and output are given in the following equations:

$$h^{\langle t \rangle} = g_1\left(W_{hh} h^{\langle t-1 \rangle} + W_{xh} x^{\langle t \rangle} + b_h\right) \quad (1)$$

$$y^{\langle t \rangle} = g_2\left(W_{yh} h^{\langle t \rangle} + b_y\right) \quad (2)$$

The total loss is the sum of the losses at time t:

$$L(\hat{y}, y) = \sum_{t=1}^{T_y} l\left(\hat{y}^{\langle t \rangle}, y^{\langle t \rangle}\right)$$

The derivative of loss with respect to the weight is then found, which by chain rule becomes $y^{<t>} = g_2(W_{yh}(g_1(W_{hh} h^{<t-1>} + W_{xh} x^{<t>} + b_h)) + b_y)$ (substituting the value of $h^{<t>}$ from equation (1) to equation (2)):

$$\frac{\partial L}{\partial W_{hh}} = \frac{\partial L^{\langle t \rangle}}{\partial \hat{y}^{\langle t \rangle}} \times \frac{\partial \hat{y}^{\langle t \rangle}}{\partial h^{\langle t \rangle}} \times \frac{\partial h^{\langle t \rangle}}{\partial W_{hh}}$$

$$(\text{Since}) \frac{\partial h^{\langle t \rangle}}{\partial W_{hh}} = \sum_{k=1}^{t} \frac{\partial h^{\langle t \rangle}}{\partial h^{\langle k \rangle}} \times \frac{\partial h^{\langle k \rangle}}{\partial W_{hh}}$$

$$(\text{and}) \frac{\partial h^{\langle t \rangle}}{\partial h^{\langle k \rangle}} = \frac{\partial h^{\langle t \rangle}}{\partial h^{\langle t-1 \rangle}} \times \frac{\partial h^{\langle t-1 \rangle}}{\partial h^{\langle t-2 \rangle}} \cdots \frac{\partial h^{\langle k+1 \rangle}}{\partial h^{\langle k \rangle}} = \prod_{j=k+1}^{t} \frac{\partial h^{\langle j \rangle}}{\partial h^{\langle j-1 \rangle}}$$

$$(\text{so}) \frac{\partial L^{\langle t \rangle}}{\partial W_{hh}} = \frac{\partial L^{\langle t \rangle}}{\partial \hat{y}^{\langle t \rangle}} \times \frac{\partial \hat{y}^{\langle t \rangle}}{\partial h^{\langle t \rangle}} \times \left(\frac{\partial h^{\langle t \rangle}}{\partial W_{hh}} + \frac{\partial h^{\langle t \rangle}}{\partial h^{\langle t-1 \rangle}} \times \frac{\partial h^{\langle t-1 \rangle}}{\partial W_{hh}} + \frac{\partial h^{\langle t \rangle}}{\partial h^{\langle t-1 \rangle}} \times \frac{\partial h^{\langle t-1 \rangle}}{\partial h^{\langle t-2 \rangle}} \times \frac{\partial h^{\langle t-2 \rangle}}{\partial W_{hh}} \cdots \right)$$

Using this formula, we can update the weights of the network. Note that if the sequence is very long, then the gradients either explode or vanish. To handle the exploding gradient, we update the gradient after k time stamps.

CHAPTER 9 RECURRENT NEURAL NETWORK

Types of RNN

An RNN can take a single input or multiple inputs and output a single vector or multiple vectors. Based on this, the RNNs can be classified into four types:

1. One to one
2. One to many
3. Many to one
4. Many to many

The **one-to-one RNN**, shown in Figure 9-5, can be perceived as a normal Neural Network. It takes an input, produces some output, and has some hidden state.

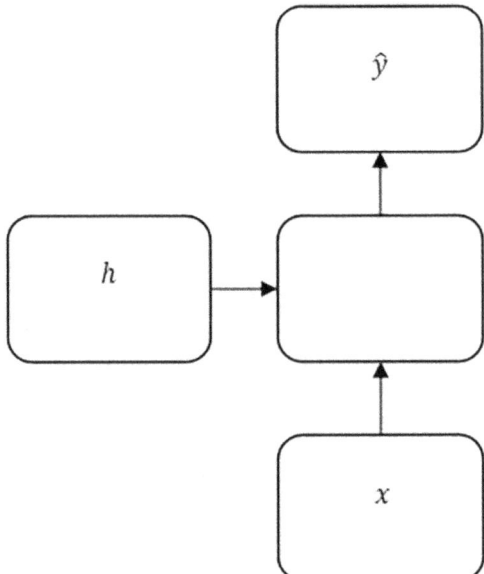

***Figure 9-5.** One-to-one RNN*

The **one-to-many RNN** shown in Figure 9-6 takes an input and produces outputs at different time stamps. Here, x is the input, $y^{<t>}$ is the output at the tth time stamp, and $h^{<t>}$ is the activation at the tth time stamp.

CHAPTER 9   RECURRENT NEURAL NETWORK

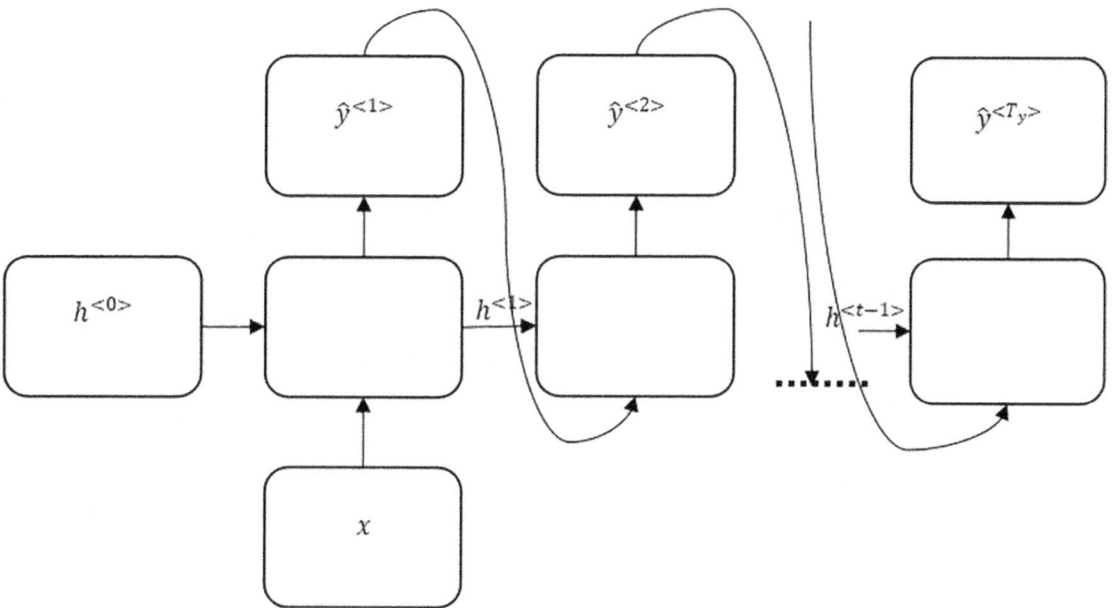

Figure 9-6. *One-to-many RNN*

These architectures are used in the following applications:

- Image captioning
- Music generation

The **many-to-one RNN** shown in Figure 9-7 takes inputs at each time stamp and produces output at the tth time stamp. Here, $x^{<t>}$ is the input, $y^{<t>}$ is the output, and $h^{<t>}$ is the activation at the tth time stamp.

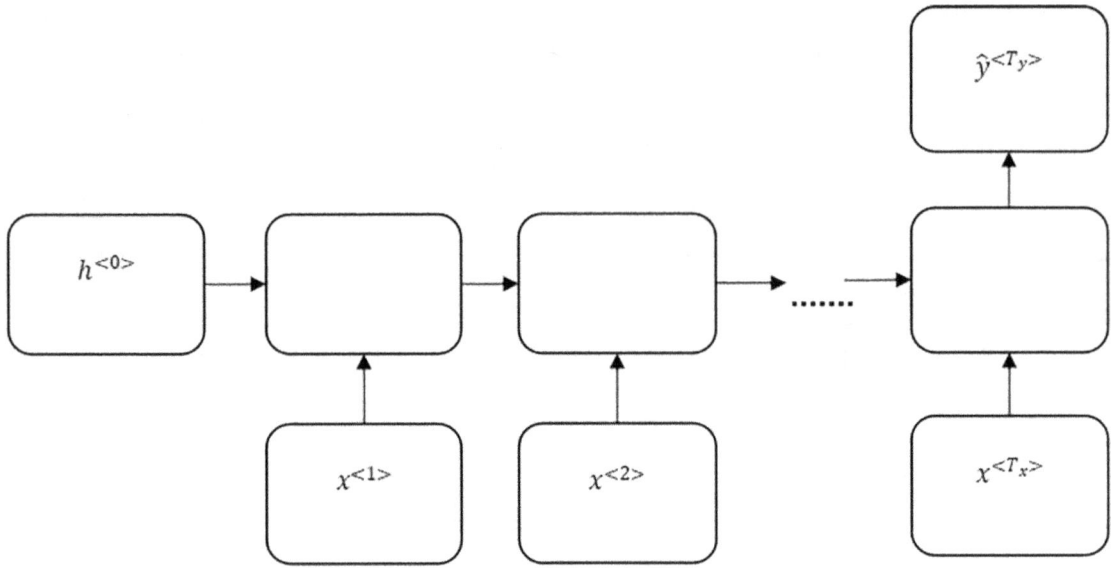

Figure 9-7. *Many-to-one RNN*

Some of the prominent applications in which these architectures are used are

- Sentiment Analysis
- Spam detection
- Stock price prediction

The **many-to-many RNN**, shown in Figure 9-8, takes inputs at different time stamps and produces outputs at the tth time stamp. Here, $x^{<t>}$ is the input, $y^{<t>}$ is the output, and $h^{<t>}$ is the activation at the tth time stamp.

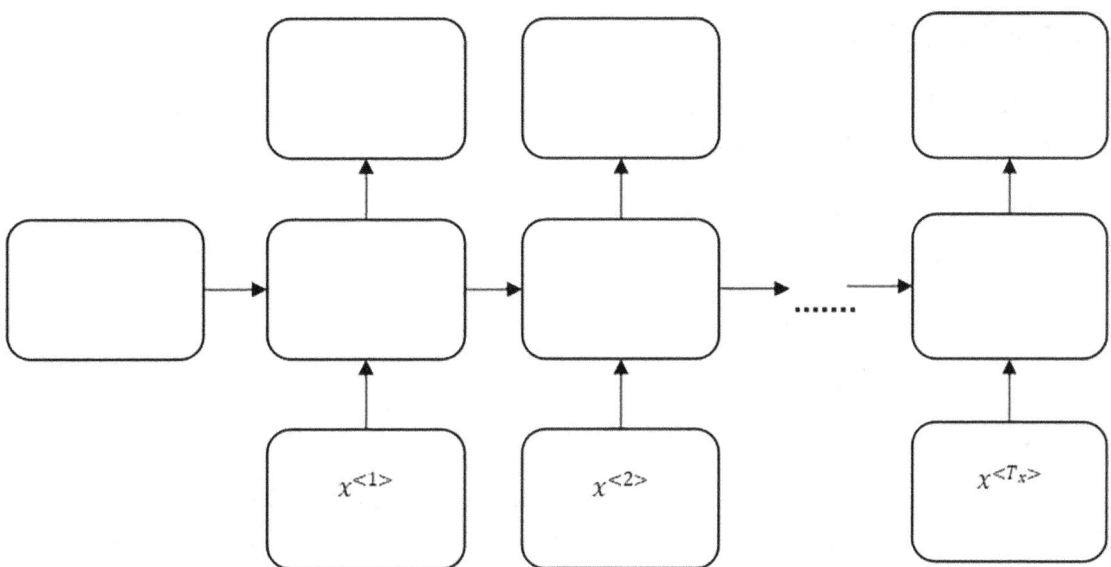

***Figure 9-8.** Many-to-many RNN like those used in parts of speech tagging*

There is another type of **many-to-many RNN** architecture that has two parts, encoder and decoder (Figure 9-9). The encoder is like a many-to-one architecture, and the decoder is one to many. In tasks like language translation, these architectures are used.

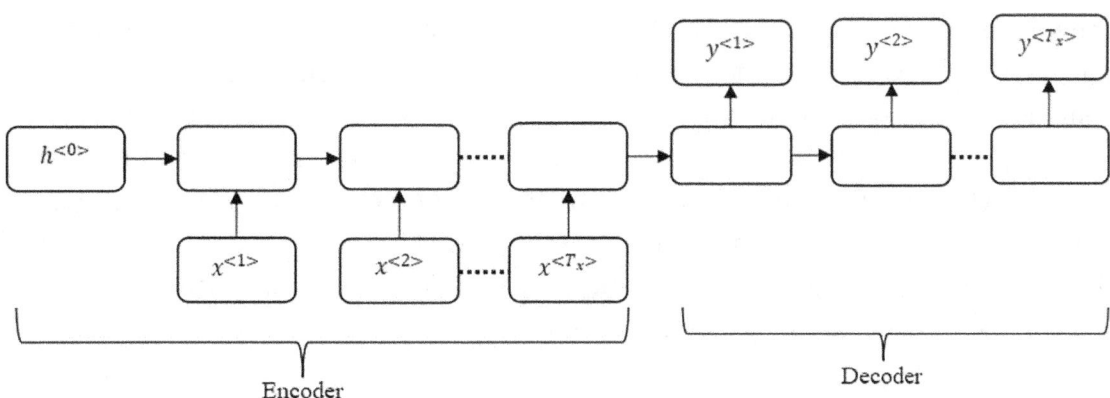

***Figure 9-9.** Many-to-many RNN: encoder and decoder type*

CHAPTER 9   RECURRENT NEURAL NETWORK

Applications

RNNs are used for sequence modeling. They (or latest sequence models) are often used to accomplish the following tasks:

- Sentiment Analysis
- Handwritten text recognition
- Image captioning
- Machine translation
- Speech-to-text conversion

to name a few. The first is an example of a many-to-one network; the second and third are examples of one-to-many models. The fourth uses the many-to-many model. The last task can be accomplished using models discussed later in this chapter. Let us explore some of these examples in detail.

Sentiment Classification

Sentiment Analysis can be implemented using a many-to-one RNN, in which the input is X (text, consisting of a sequence of words) and the output is an integer y representing the sentiment. Here the length of X is the same as the length of a sentence. However, the length of each sentence may not be the same, so we consider a maximum length, and the sentences that do not have that many words are padded with zeros or fixed numbers.

The problem is now to convert the words of a sentence into embeddings. Consider each word being represented as an embedding of m numbers. If the maximum length of sentences is considered to be n, then a sentence will be represented as a 2D array of dimension $n \times m$. So, in each iteration, the model is trained using sentence X, $x_i \in X$ and $x_i \in R^m$, and $y \in (0, 1)$ in the case of binary classification or else equal to the number of sentiments.

The following Listing 9-1 classifies the given sentences into positive or negative sentiments. The four different RNN models were created and evaluated on the IMDB movie review sentiment dataset. This dataset contains 50,000 movie reviews, evenly split into positive and negative sentiments. The dataset was preprocessed by removing stop words, tokenizing, and padding the reviews. The four models were created including a simple RNN with a single layer having 32 units, a stacked RNN with two layers having 32 units in each layer, a bidirectional RNN with a single layer having 32 units, and a stacked bidirectional RNN with two layers having 32 units in each layer. The variation of accuracy and loss with the number of epochs is shown in figures from Figure 9-10 to Figure 9-13. The mean validation accuracy for each model was calculated and is shown in Table 9-1.

Listing 9-1. Sentiment classification using the IMDB dataset

Code:

```
# 1. Import the IMDB dataset from tensorflow.keras.datasets, stopwords
from nltk.corpus. The tensorflow.keras.models, tensorflow.keras.layers
are imported to design a sequential model having Embedding, RNN, and
Bidirectional layers.
import numpy as np
from tensorflow.keras.datasets importimdb
from tensorflow.keras.preprocessing.sequence import pad_sequences
from tensorflow.keras.preprocessing.text import Tokenizer
from nltk.corpus import stopwords
import nltk
from tensorflow.keras.models import Sequential
from tensorflow.keras.layers import Embedding, SimpleRNN, Dense, Bidirectional
from matplotlib import pyplot as plt
# 2. The stopwords are downloaded from NLTK
nltk.download('stopwords')
# 3. The IMDB dataset is downloaded and limited to the top 10,000 most
frequent words.
max_features = 10000
(X_train, y_train), (X_test, y_test) = imdb.load_data(num_words=max_features)
# 4. Create a reverse dictionary to decode reviews back to words
word_index = imdb.get_word_index()
reverse_word_index = dict([(value, key) for (key, value) in word_index.items()])
```

```python
# 5. Create a function to decode reviews from sequences of integers
to words
def decode_review(encoded_review):
    return ' '.join([reverse_word_index.get(i - 3, '?') for i in encoded_
    review])
# 6. Decode all reviews in the training and test sets
decoded_train = [decode_review(review) for review in X_train]
decoded_test = [decode_review(review) for review in X_test]
# 7. Remove the stop words from the reviews
stop_words = set(stopwords.words('english'))
def remove_stop_words(text):
    return ' '.join([word for word in text.split() if word not in
    stop_words])
cleaned_train = [remove_stop_words(review) for review in decoded_train]
cleaned_test = [remove_stop_words(review) for review in decoded_test]
# 8. Tokenize the cleaned reviews using the Tokenizer function imported
from tensorflow.keras.preprocessing.text
tokenizer = Tokenizer(num_words=max_features)
tokenizer.fit_on_texts(cleaned_train)
# 9. Convert the tokenized reviews to sequences
train_sequences = tokenizer.texts_to_sequences(cleaned_train)
test_sequences = tokenizer.texts_to_sequences(cleaned_test)
# 10. Pad the sequences to ensure they all have the same length
maxlen = 100
X_train = pad_sequences(train_sequences, maxlen=maxlen)
X_test = pad_sequences(test_sequences, maxlen=maxlen)
# 11. Create a function to create, compile, and train a model
def compile_and_train(model, epochs=10):
    model.compile(optimizer='adam', loss='binary_crossentropy',
    metrics=['acc'])
    history = model.fit(X_train, y_train, epochs=epochs, batch_
    size=32,validation_split=0.3)
    return history
```

12. Create a functionto plot the accuracy and loss curves from the history obtained of the trained model.
```
def plot_history(history, title):
    plt.figure(figsize=(12, 6))
    plt.plot(history.history['acc'], label='Train Accuracy')
    plt.plot(history.history['val_acc'], label='Validation Accuracy')
    plt.title(f'{title} Accuracy')
    plt.xlabel('Epochs')
    plt.ylabel('Accuracy')
    plt.legend()
    plt.show()
    plt.figure(figsize=(12, 6))
    plt.plot(history.history['loss'], label='Train Loss')
    plt.plot(history.history['val_loss'], label='Validation Loss')
    plt.title(f'{title} Loss')
    plt.xlabel('Epochs')
    plt.ylabel('Loss')
    plt.legend()
    plt.show()
```
13. Model1
```
model_1 = Sequential()
model_1.add(Embedding(max_features, 32))
model_1.add(SimpleRNN(32))
model_1.add(Dense(1, activation='sigmoid'))
history_1 = compile_and_train(model_1)
plot_history(history_1, 'Simple RNN')
```
14. Model 2
```
model_2 = Sequential()
model_2.add(Embedding(max_features, 32))
model_2.add(SimpleRNN(32, return_sequences=True))
model_2.add(SimpleRNN(32))
model_2.add(Dense(1, activation='sigmoid'))
history_2 = compile_and_train(model_2)
plot_history(history_2, 'Stacked Simple RNN')
```

15. Model3
```
model_3 = Sequential()
model_3.add(Embedding(max_features, 32))
model_3.add(Bidirectional(SimpleRNN(32)))
model_3.add(Dense(1, activation='sigmoid'))
history_3 = compile_and_train(model_3)
plot_history(history_3, 'Bidirectional Simple RNN')
```
16. Model 4
```
model_4 = Sequential()
model_4.add(Embedding(max_features, 32))
model_4.add(Bidirectional(SimpleRNN(32, return_sequences=True)))
model_4.add(Bidirectional(SimpleRNN(32)))
model_4.add(Dense(1, activation='sigmoid'))
history_4 = compile_and_train(model_4)
plot_history(history_4, 'Stacked Bidirectional Simple RNN')
```
17. Calculate the mean validation accuracy for each model
```
mean_accuracies = {
    'Simple RNN': np.mean(history_1.history['val_acc']),
    'Stacked Simple RNN': np.mean(history_2.history['val_acc']),
    'Bidirectional Simple RNN': np.mean(history_3.history['val_acc']),
    'Stacked Bidirectional Simple RNN': np.mean(history_4.
    history['val_acc'])
}
```
18. Print the mean validation accuracy for each model
```
for model_name, mean_acc in mean_accuracies.items():
    print(f"{model_name} mean validation accuracy: {mean_acc:.4f}")
```

CHAPTER 9   RECURRENT NEURAL NETWORK

Output:

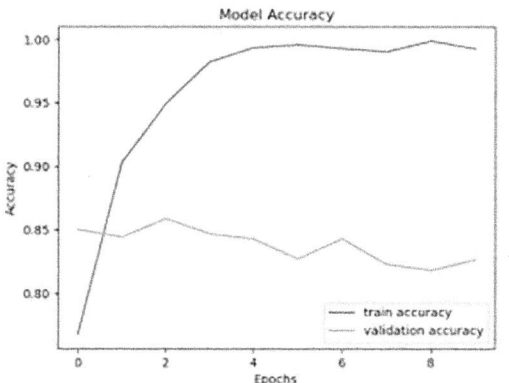 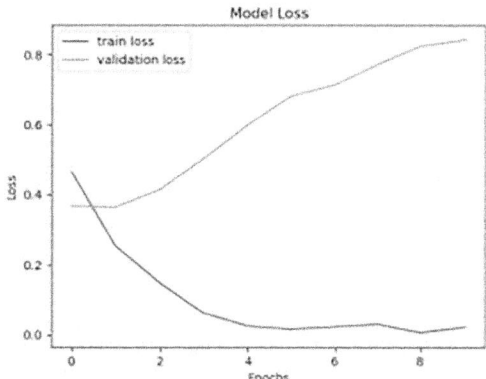

Figure 9-10. *Loss and accuracy curves: Model 1*

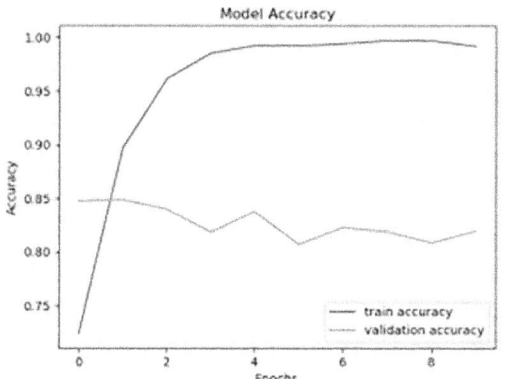 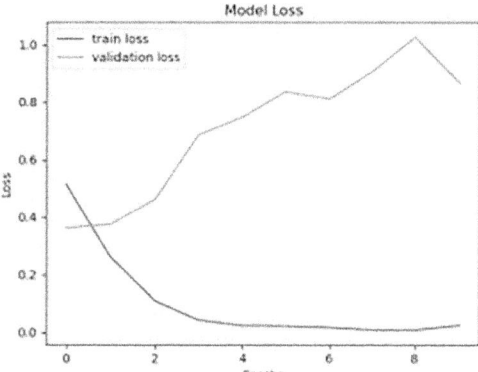

Figure 9-11. *Loss and accuracy curves: Model 2*

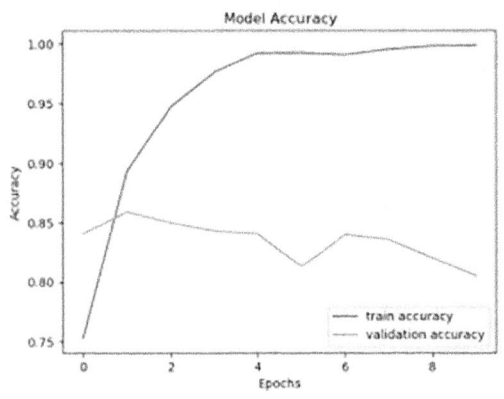
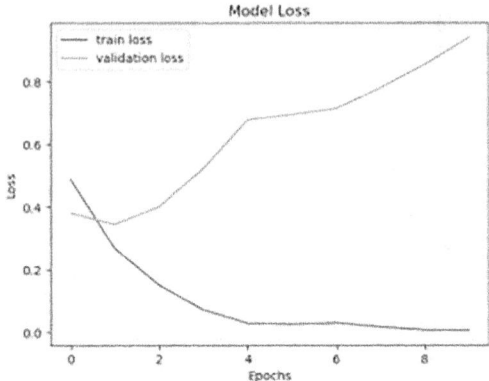

Figure 9-12. Loss and accuracy curves: Model 3

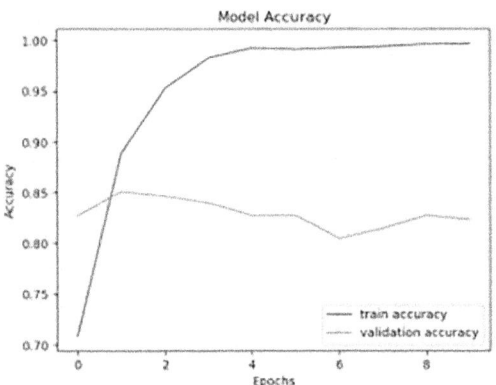
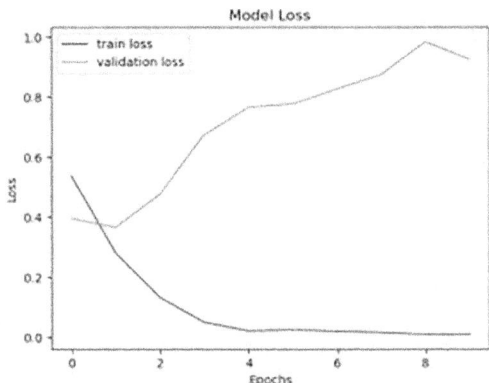

Figure 9-13. Loss and accuracy curves: Model 4

The results of the above experiments are summarized in Table 9-1.

Table 9-1. Mean Validation Accuracy of Four Different RNN Models on the IMDB Dataset

Architecture	Mean Validation Accuracy
Simple RNN with a single layer having 32 units	0.8325
Stacked RNN with two layers having 32 units in each layer	0.8173
Bidirectional RNN with a single layer having 32 units	0.8323
Stacked bidirectional RNN with two layers having 32 units in each layer	0.8031

Having seen an application of a many-to-one model, let us have a look at an application of many-to-many models.

Parts of Speech Tagging

Parts of speech (POS) tagging maps each word in a sentence to the corresponding part of speech. It can be implemented using a many-to-many RNN model in which the input is a sentence and the output is a number corresponding to each part of speech. Consider the following sentence:

"Nishant is traveling to the United States to pursue a postgraduate degree."

The parts of speech corresponding to each word in the above sentence are as follows:

- Nishant: Noun
- is: Verb
- traveling: Verb
- to: Preposition
- the: Determiner
- United States: Noun
- to: Preposition
- pursue: Verb
- a: Determiner
- postgraduate: Noun
- degree: Noun

The problem is now to convert the words of a sentence into embeddings. Consider each word being represented as an embedding of m numbers, and if the maximum length of sentences is considered to be n, then a sentence is represented as a 2D array of dimension $n \times m$. So, in each iteration, the model is trained using sentence X, $x_i \in X$ and $x_i \in R^m$, $y \in (1, 2, 3,,...)$ equal to the number of parts of speech.

The Penn Treebank dataset provided by NLTK consists of tagged sentences in English and contains over 4.5 million words of American English text, taken from a variety of sources. The dataset was preprocessed by extracting unique words and tags, mapping them to indices, and converting sentences to sequences of word indices and

CHAPTER 9   RECURRENT NEURAL NETWORK

corresponding tag indices. The four different RNN architectures were implemented (Listing 9-2) using **Keras**. **Model 1** utilized a single-layer simple RNN with 64 units. **Model 2** employed a stacked simple RNN with two layers, each containing 64 units. **Model 3** employed a single-layer bidirectional RNN with 64 units. **Model 4** incorporated a stacked bidirectional RNN with two layers, each containing 64 units. Each model was trained for ten epochs with a batch size of 32. The accuracy and loss curves for each model are shown in Figure 9-14 to Figure 9-17. The mean validation accuracies were computed for each model architecture and are shown in Table 9-2.

Listing 9-2. POS tagging using the Treebank dataset

Code:
#1. Import the treebank dataset from *nltk.corpus*. The *tensorflow.keras. models*, *tensorflow.keras.layers* are imported to design a sequential model having Embedding, RNN, Bidirectional, and TimeDistributed layers.

```
import numpy as np
import tensorflow as tf
from tensorflow.keras.preprocessing.sequence import pad_sequences
from tensorflow.keras.utils import to_categorical
from tensorflow.keras.models import Sequential
from tensorflow.keras.layers import Embedding, SimpleRNN, Dense, TimeDistributed, Bidirectional
import matplotlib.pyplot as plt
import nltk
from sklearn.model_selection import train_test_split
from nltk.corpus import treebank
nltk.download('treebank')
```

#2. Create a function to load the treebank dataset

```
def load_data():
    sentences = treebank.tagged_sents()
    return sentences
```

#3. Create a function to prepare the data by creating dictionaries for word-to-index and tag-to-index mappings

```
def preprocess_data(sentences):
    words = set()
    tags = set()
```

```
        for sentence in sentences:
            for word, tag in sentence:
words.add(word)
tags.add(tag)
    word2idx = {w: i + 2 for i, w in enumerate(words)}
    word2idx["PAD"] = 0  # Padding token
    word2idx["UNK"] = 1  # Unknown token
    tag2idx = {t: i + 1 for i, t in enumerate(tags)}
    tag2idx["PAD"] = 0  # Padding tag
    idx2word = {i: w for w, i in word2idx.items()}
    idx2tag = {i: t for t, i in tag2idx.items()}
    return word2idx, tag2idx, idx2word, idx2tag
```

#4. **Load and pre-process the dataset using the above functions**

```
sentences = load_data()
word2idx, tag2idx, idx2word, idx2tag = preprocess_data(sentences)
```

#5. **Create a function to convert sentences to sequences of indices**

```
def convert_sentences_to_sequences(sentences, word2idx, tag2idx):
    X = [[word2idx.get(word, word2idx["UNK"]) for word, _ in sentence] for
        sentence in sentences]
    y = [[tag2idx[tag] for _, tag in sentence] for sentence in sentences]
    return X, y
```

#6. **Convert the sentences to padded sequences and one-hot encoded labels**

```
X, y = convert_sentences_to_sequences(sentences, word2idx, tag2idx)
max_len = 50  # Maximum sequence length
X = pad_sequences(X, maxlen=max_len, padding="post")
y = pad_sequences(y, maxlen=max_len, padding="post")
y = [to_categorical(i, num_classes=len(tag2idx)) for i in y]
```

#7. **Split the dataset into training and test sets**

```
X_train, X_test, y_train, y_test = train_test_split(X, y, test_size=0.1)
```

#8. **Model 1**

```
model_1 = Sequential()
model_1.add(Embedding(input_dim=len(word2idx), output_dim=64, input_length=max_len))
model_1.add(SimpleRNN(units=64, return_sequences=True, recurrent_dropout=0.1))
```

CHAPTER 9   RECURRENT NEURAL NETWORK

```
model_1.add(TimeDistributed(Dense(len(tag2idx), activation="softmax")))
model_1.compile(optimizer="adam", loss="categorical_crossentropy",
metrics=["accuracy"])
history_1 = model_1.fit(X_train, np.array(y_train), batch_size=32,
epochs=5, validation_data=(X_test, np.array(y_test)), verbose=1)
```
#9. Model 2
```
model_2 = Sequential()
model_2.add(Embedding(input_dim=len(word2idx), output_dim=64, input_
length=max_len))
model_2.add(SimpleRNN(units=64, return_sequences=True, recurrent_
dropout=0.1))
model_2.add(SimpleRNN(units=64, return_sequences=True, recurrent_
dropout=0.1))
model_2.add(TimeDistributed(Dense(len(tag2idx), activation="softmax")))
model_2.compile(optimizer="adam", loss="categorical_crossentropy",
metrics=["accuracy"])
history_2 = model_2.fit(X_train, np.array(y_train), batch_size=32,
epochs=5, validation_data=(X_test, np.array(y_test)), verbose=1)
```
#10. Model 3
```
model_3 = Sequential()
model_3.add(Embedding(input_dim=len(word2idx), output_dim=64, input_
length=max_len))
model_3.add(Bidirectional(SimpleRNN(units=64, return_sequences=True,
recurrent_dropout=0.1)))
model_3.add(TimeDistributed(Dense(len(tag2idx), activation="softmax")))
model_3.compile(optimizer="adam", loss="categorical_crossentropy",
metrics=["accuracy"])
history_3 = model_3.fit(X_train, np.array(y_train), batch_size=32,
epochs=5, validation_data=(X_test, np.array(y_test)), verbose=1)
```
#11. Model 4
```
model_4 = Sequential()
model_4.add(Embedding(input_dim=len(word2idx), output_dim=64, input_
length=max_len))
model_4.add(Bidirectional(SimpleRNN(units=64, return_sequences=True,
recurrent_dropout=0.1)))
```

```python
model_4.add(Bidirectional(SimpleRNN(units=64, return_sequences=True,
recurrent_dropout=0.1)))
model_4.add(TimeDistributed(Dense(len(tag2idx), activation="softmax")))
model_4.compile(optimizer="adam", loss="categorical_crossentropy",
metrics=["accuracy"])
history_4 = model_4.fit(X_train, np.array(y_train), batch_size=32,
epochs=5, validation_data=(X_test, np.array(y_test)), verbose=1)
```

#12. **Create a function to plot accuracy and loss curves from the history obtained of each trained model**

```python
def plot_history(history, model_name):
    plt.figure(figsize=(12, 6))
    plt.subplot(1, 2, 1)
    plt.plot(history.history['accuracy'])
    plt.plot(history.history['val_accuracy'])
    plt.title(f'{model_name} Model Accuracy')
    plt.xlabel('Epoch')
    plt.ylabel('Accuracy')
    plt.legend(['Train', 'Val'], loc='upper left')
    plt.subplot(1, 2, 2)
    plt.plot(history.history['loss'])
    plt.plot(history.history['val_loss'])
    plt.title(f'{model_name} Model Loss')
    plt.xlabel('Epoch')
    plt.ylabel('Loss')
    plt.legend(['Train', 'Val'], loc='upper left')
    plt.tight_layout()
    plt.show()
```

#13. **Plot the accuracy and loss curves for each model using the above function**

```python
plot_history(history_1, "Model 1")
plot_history(history_2, "Model 2")
plot_history(history_3, "Model 3")
plot_history(history_4, "Model 4")
```

#14. **Create a function to calculate mean validation accuracy**

```python
def mean_validation_accuracy(history):
```

CHAPTER 9 RECURRENT NEURAL NETWORK

```
    val_acc = history.history['val_accuracy']
    mean_acc = np.mean(val_acc)
    return mean_acc
#15. Compute the mean validation accuracy for each model
mean_acc_1 = mean_validation_accuracy(history_1)
mean_acc_2 = mean_validation_accuracy(history_2)
mean_acc_3 = mean_validation_accuracy(history_3)
mean_acc_4 = mean_validation_accuracy(history_4)
```

Output:

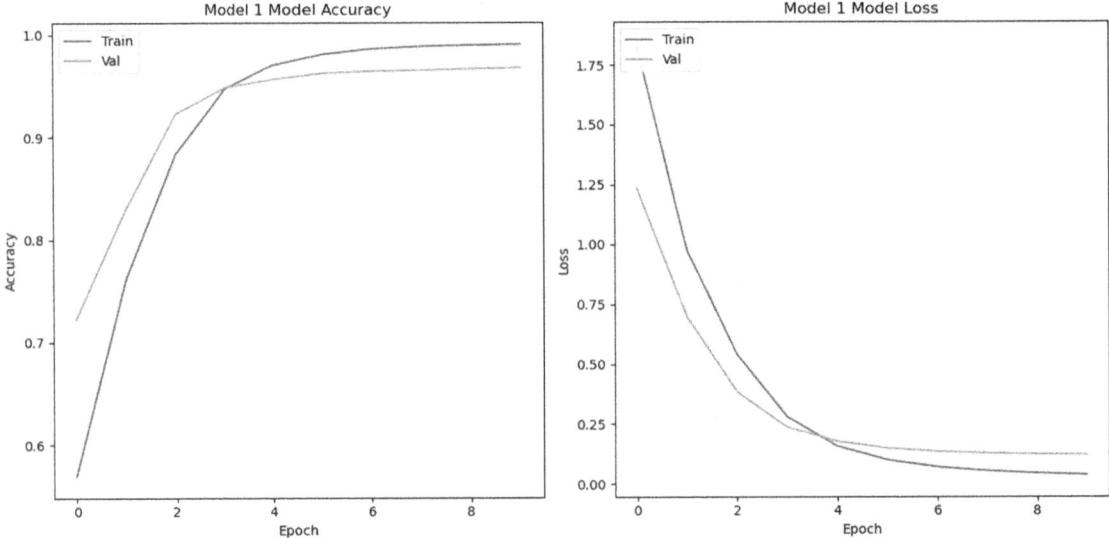

Figure 9-14. *Loss and accuracy curves: Model 1*

CHAPTER 9 RECURRENT NEURAL NETWORK

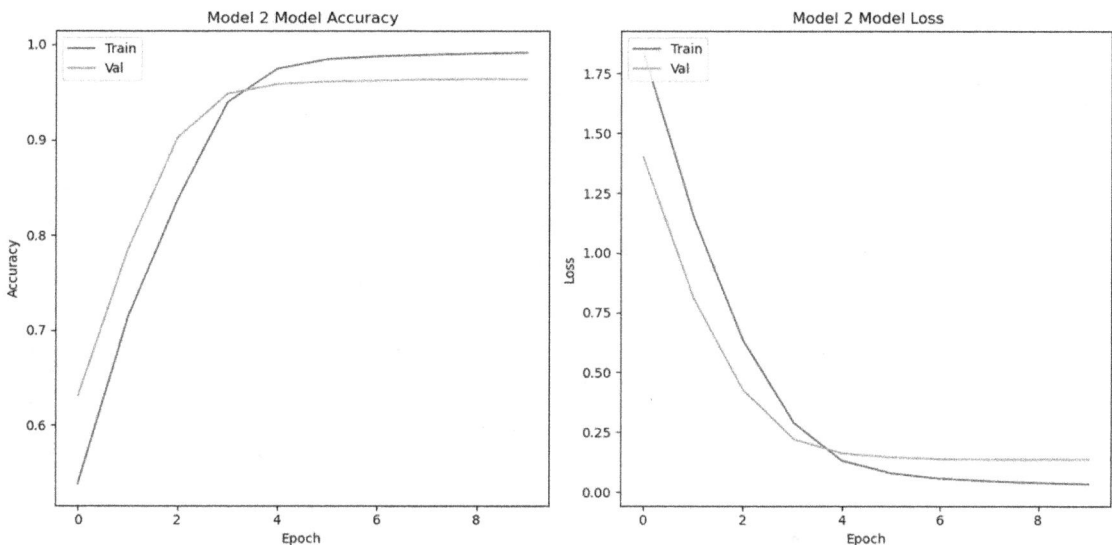

Figure 9-15. *Loss and accuracy curves: Model 2*

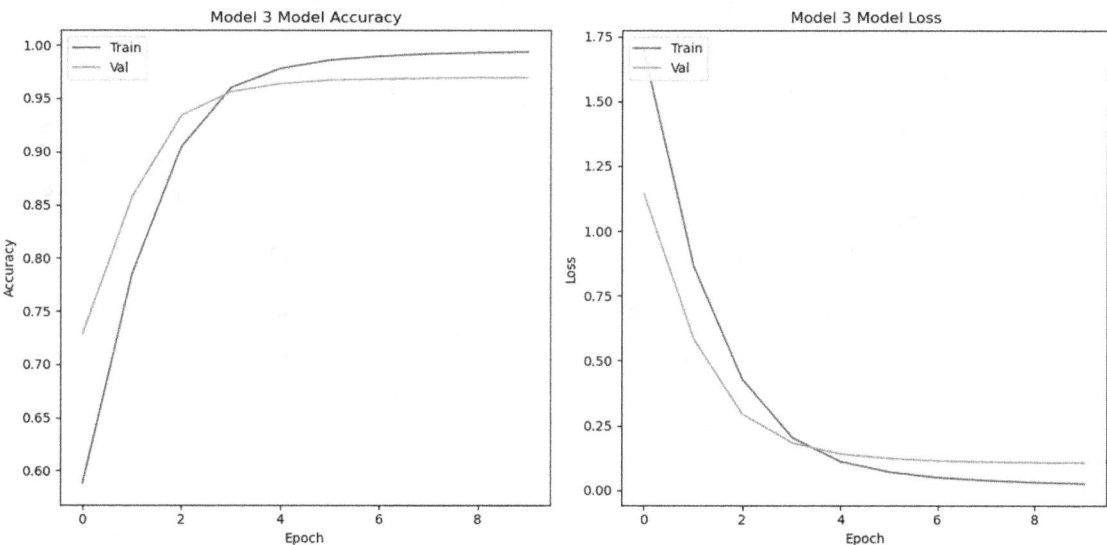

Figure 9-16. *Loss and accuracy curves: Model 3*

247

CHAPTER 9 RECURRENT NEURAL NETWORK

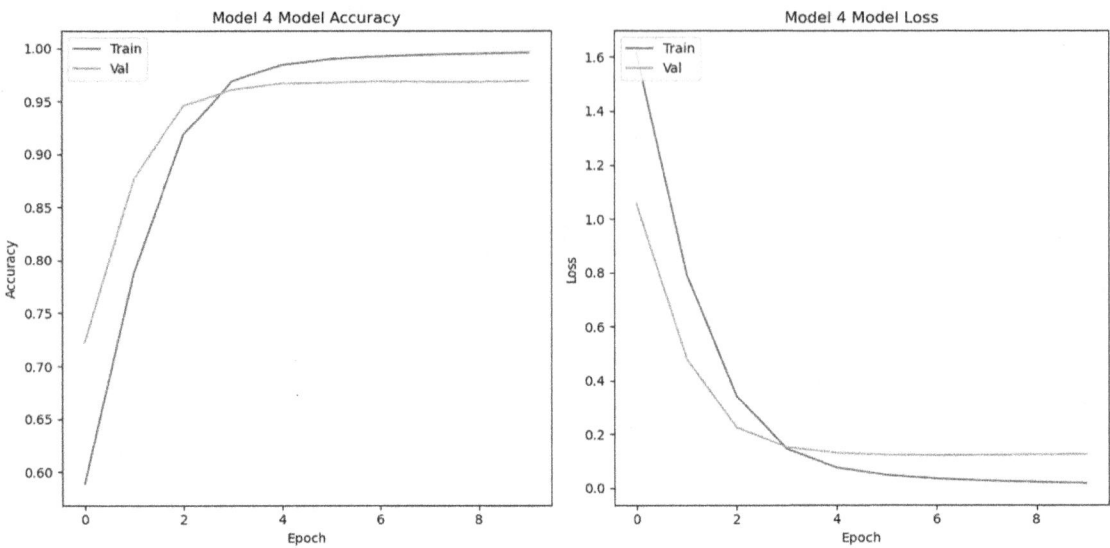

Figure 9-17. *Loss and accuracy curves: Model 4*

The results of the above experiments are summarized in Table 9-2.

Table 9-2. *Mean Validation Accuracy of Four Different RNN Models on the Treebank Dataset*

Architecture	Mean Validation Accuracy
Simple RNN with a single layer having 64 units	0.9206
Stacked RNN with two layers having 64 units in each layer	0.9040
Bidirectional RNN with a single layer having 64 units	0.9284
Stacked bidirectional RNN with two layers having 64 units in each layer	0.9314

Note that the bidirectional RNN performs better as it can capture both the forward and the backward context. That is, it finds the relation of an element with the element before it and those after it. Let us have a look at an application that uses a one-to-many RNN model.

Handwritten Text Recognition

You are given images containing some handwritten text in English, and you are required to obtain the text corresponding to it. That is, you are required to recognize images of handwritten text. How do you think you could solve the problem?

One of the simplest solutions to this problem, based on what we have studied so far, is to convert the input pictures to embeddings using a CNN model and then give this as input to a one-to-many RNN model as shown in Figure 9-18.

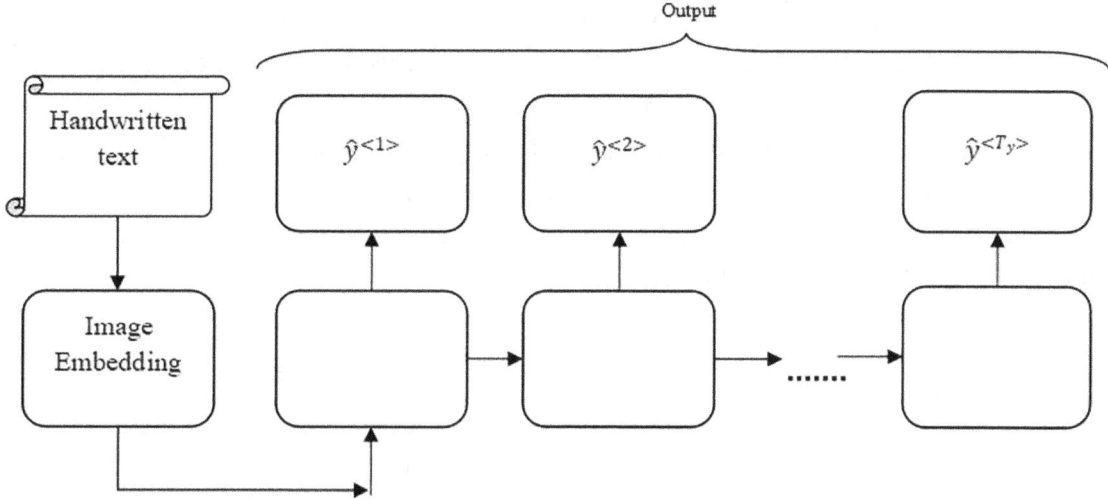

Figure 9-18. *Handwritten text recognition model*

To accomplish this task, you can try the following:

a) Create embedding of the input image using some pretrained Convolutional Neural Network.

b) Create embeddings using autoencoders (Chapter 11).

c) Use RNN with a single layer having 64 units (you can change the number of units if you want).

d) Use RNN with two layers, having 64 and 32 units.

e) Use dropout and analyze the effect of introducing this layer on the performance of the model with the test set.

CHAPTER 9 RECURRENT NEURAL NETWORK

The reader is expected to try all the combinations of the above and find the model that works well. You can obtain any publicly available dataset to accomplish the task. One of the options is as follows:

```
https://www.kaggle.com/datasets/landlord/handwriting-recognition
```

Speech to Text

You are given audio containing the recordings of some hours of speech in English and the corresponding transcriptions. You are required to get the transcript corresponding to yet unheard (by the model) speech. That is, you are required to transcribe speech. How do you think you could solve the problem?

Again, there can be many interesting solutions to this problem, one of which, based on what we have studied so far, is to obtain the embeddings of the audio data (try obtaining the embeddings of segments first) and then give these as input to a one-to-many RNN model as shown in Figure 9-19.

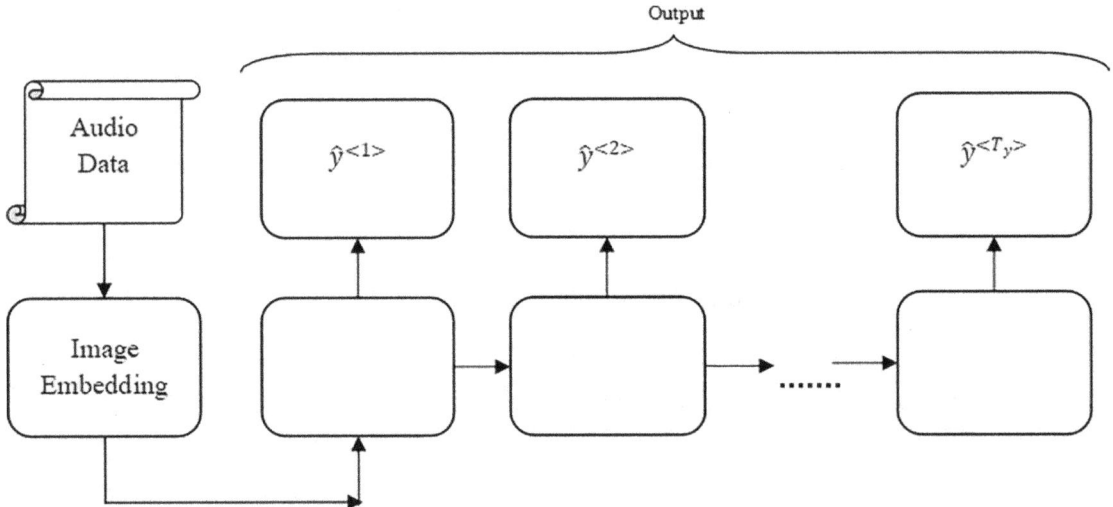

Figure 9-19. *Speech to text*

To accomplish this task, you can try the following:

a) Create embeddings of the given audio data using Mel-spectrograms or Cepstral, followed by the application of Local Binary Pattern.

b) Create embeddings of the images obtained in (a) using autoencoders (Chapter 11).

c) Use RNN with a single layer having 32 units (you can change the number of units if you want).

d) Use RNN with two layers, having 32 and 16 units.

e) Use dropout and analyze the effect of introducing this layer on the performance of the model with the test set.

The reader is expected to try all the combinations of the above and find the model that works well. You can obtain any publicly available dataset to accomplish the task. One of the options is as follows:

https://www.openslr.org/12

Conclusion

This chapter introduced the Recurrent Neural Network, a sequence model capable of handling sequence data. The architectures of RNN and the algorithm to train the model are discussed in the chapter. This chapter contains some very interesting applications of RNN including Sentiment Analysis, parts of speech tagging, and handwritten text recognition. The reader is expected to attempt the exercises to get hold of the concepts studied in this chapter. The next chapter takes the discussion forward and introduces Gate Recurrent Unit (GRU) and Long Short-Term Memory (LSTM) that handle the problem of vanishing gradient gracefully.

Exercises
Multiple-Choice Questions

1. What type of data is RNN used to handle?

    a. Imaging data

    b. Sequential data

    c. Numeric tabular data

    d. Graph data

CHAPTER 9   RECURRENT NEURAL NETWORK

2. In Neural Networks, how are the input and output related?

   a. They are linearly dependent on each other.

   b. They are independent of each other.

   c. They are processed sequentially.

   d. They are processed recursively.

3. How does an RNN process information vis-à-vis a general Neural Network?

   a. Independently

   b. Randomly

   c. In parallel

   d. Sequentially

4. Which of the following is true regarding the parameters in RNNs?

   a. Initialized randomly.

   b. Share parameters across each layer.

   c. Different parameters for each layer.

   d. They do not use parameters.

5. Which algorithm do RNNs use to compute the loss?

   a. Gradient descent

   b. Backpropagation

   c. BPTT

   d. Genetic algorithm

6. How do traditional feed-forward networks differ from RNNs in terms of weights?

   a. Feed-forward networks share the same weights across each layer.

   b. They have different weights for each layer.

   c. Feed-forward networks have different weights for each layer.

   d. Both have the same weights across each layer.

CHAPTER 9   RECURRENT NEURAL NETWORK

7. What can RNNs handle that traditional feed-forward networks cannot?

    a. Fixed-length input data

    b. Sequential data

    c. Input data of any length

    d. Non-sequential data

8. Which of the following activation functions is commonly used in RNNs?

    a. Softmax

    b. Tanh

    c. Leaky ReLU

    d. SIREN

9. What challenges do RNNs face when capturing long-term dependencies?

    a. Overfitting

    b. Underfitting

    c. Exponential increase or decrease in multiplicative gradients

    d. Lack of enough training data

10. What happens during the BPTT process in RNNs?

    a. The network unfolds in multiple layers.

    b. The network computes gradients at each time step independently.

    c. The network unfolds in multiple time steps and computes gradients over these steps.

    d. The network uses a genetic algorithm to update parameters.

CHAPTER 9 RECURRENT NEURAL NETWORK

Theory

1. Why are RNNs better compared with Neural Networks for sequential data?

2. Explain Backpropagation Through Time.

3. What are the various types of RNNs? Give examples of each type.

Image Captioning

You are given images along with their captions, and you are required to obtain the caption corresponding to a new image. How do you think you could solve the problem?

Hint: One of the simplest solutions to this problem, based on what we have studied so far, is to convert the input pictures to embeddings using a CNN model and then give this as input to a one-to-many RNN model as shown in Figure 9-20.

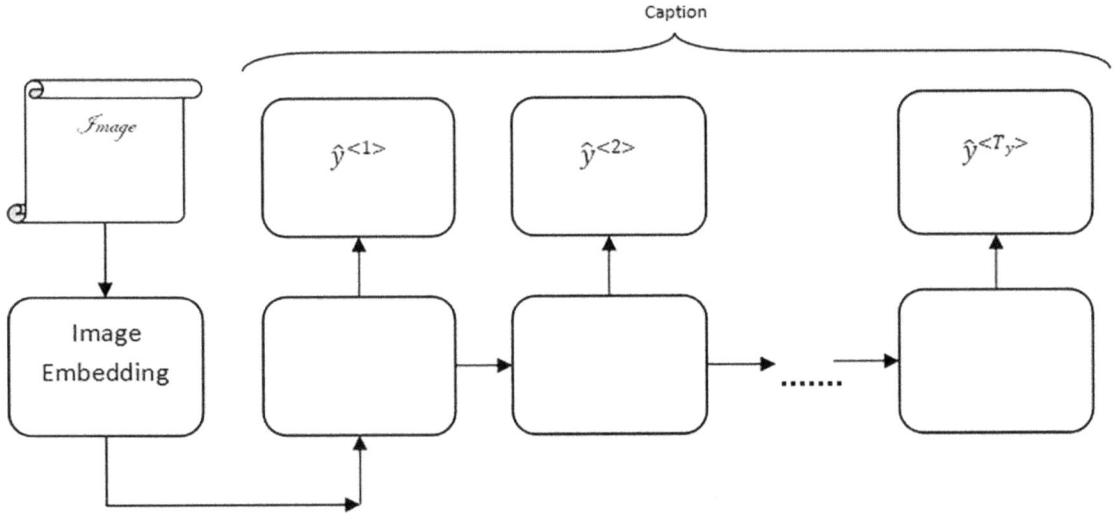

Figure 9-20. *Image captioning*

To accomplish this task, you can try the following:

a) Create embeddings of the given image using a pretrained Convolutional Neural Network such as VGG 19.

b) Create embeddings of the given image using autoencoders (Chapter 11).

CHAPTER 9   RECURRENT NEURAL NETWORK

c) Use RNN with a single layer having 64 units.

d) Use RNN with two layers, having 64 and 32 units.

e) Use dropout and analyze the effect of introducing this layer on the performance of the model with the test set.

The reader is expected to try all the combinations of the above and find the model that works well. You can obtain any publicly available dataset to accomplish the task. One of the options is as follows:

https://paperswithcode.com/dataset/conceptual-captions

References

[1] Zemel, R., Martens, J. and Sutskever, I. COMS 4995 Lecture 8: Recurrent Neural Networks. In *COMS 4995 Lecture 8: Recurrent Neural Networks* (pp. 1–34) (2011). https://www.cs.columbia.edu/~zemel/Class/Nndl/files/lec08.pdf

[2] Hinton, G. (n.d.). *CSC2535 2013: Advanced Machine Learning Lecture 10 Recurrent Neural Networks Getting targets when modeling sequences.* https://www.cs.toronto.edu/~hinton/csc2535/notes/lec10new.pdf

[3] Recurrent neural networks. In *MIT 6.036 Fall 2019* (2019). https://openlearninglibrary.mit.edu/assets/courseware/v1/0de27572f5d771b35ad094df49a8e200/asset-v1:MITx+6.036+1T2019+type@asset+block/notes_chapter_Recurrent_Neural_Networks.pdf

[4] Li, F.-F., Krishna, R. & Xu, D. (2021). *Lecture 10.* https://cs231n.stanford.edu/slides/2021/lecture_10.pdf

CHAPTER 10

Gated Recurrent Unit and Long Short-Term Memory

Introduction

So far, we have studied Dense Neural Networks (DNNs) and various optimization techniques. We also studied the Convolutional Neural Networks (CNNs) that can handle imaging data and the Recurrent Neural Networks (RNNs) capable of handling sequence data. Let us pause for a moment and explore sequence data from another perspective.

Consider a program on a popular TV channel in Shangri La, hosted by their star anchor Mr. A. He only talks about four things:

  a) His favorite leader
  b) Why some people are problematic
  c) Advantages of all the policies of dispensation
  d) The mistakes of the previous dispensations of the country

In order to guess what his today's topic will be, you create a Neural Network. The inputs to the network are

  i) Day of the week (number from 1 to 7)
  ii) Whether a new policy is announced that day (0 or 1)
  iii) Whether elections are approaching or going on (0 or 1)

CHAPTER 10 GATED RECURRENT UNIT AND LONG SHORT-TERM MEMORY

You train the network using historical data and try to predict today's topic. However, your network does not predict effectively. Now you realize that there is a sequence in Mr. A's deciding the topic. A particular topic always comes after one topic. To handle such sequential data, you design a network that takes input and predicted output as the input in the next time stamp. This is called a recursive network (Figure 10-1).

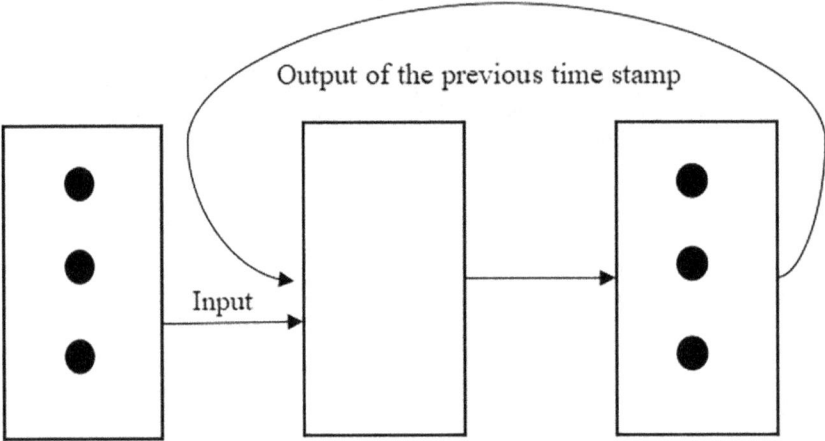

Figure 10-1. *Recurrent unit*

However, there is a problem with such kind of architecture. For the sake of simplification, assume that a single scalar is output through the network. If that scalar is greater than one, then at one point in time the output will explode or become very large, whereas if it is less than one, then after successive multiplications its effect will become negligible. The first problem was discussed in the previous chapter, and the second problem is referred to as the vanishing gradient, which can be handled using two models: Long Short-Term Memory (LSTM) and Gated Recurrent Unit (GRU), discussed in this chapter. Let us begin the discussion with GRU.

GRU

GRU is a type of RNN that can gracefully handle the problem of vanishing gradient. The hidden state of GRU depends on the previous hidden state h_{t-1} and the new memory. Assume that z_t is a scalar between 0 and 1; then h_t can be found using the following equation:

$$h_t = (1-z_t) * \widetilde{h_t} + z_t * h_{t-1}$$

where h_t is the new memory and z_t is the factor that controls what part of the previous hidden state goes into the new hidden state. Here "*" represents point-by-point multiplication.

z_t, referred to as the update gate, is the combination of x_t and h_{t-1} given as follows:

$$z_t = \sigma\left(W_{xz}x_t + W_{hz}h_{t-1}\right)$$

Now, r_t (reset gate) is also calculated as a combination of x_t and h_{t-1}. It tells us how much part of h_{t-1} is summated to the new memory state:

$$r_t = \sigma\left(W_{xr}x_t + W_{hr}h_{t-1}\right)$$

The new memory is calculated as

$$\widetilde{h}_t = tanh\left(W_h\left(r_t * h_{t-1}\right) + W_x x_t\right)$$

Note that if

$$z_t = 1 \text{ then } h_t = h_{t-1}$$

$$z_t = 0 \text{ then } h_t = \widetilde{h}_t$$

The above process can be depicted as follows (Figure 10-2).

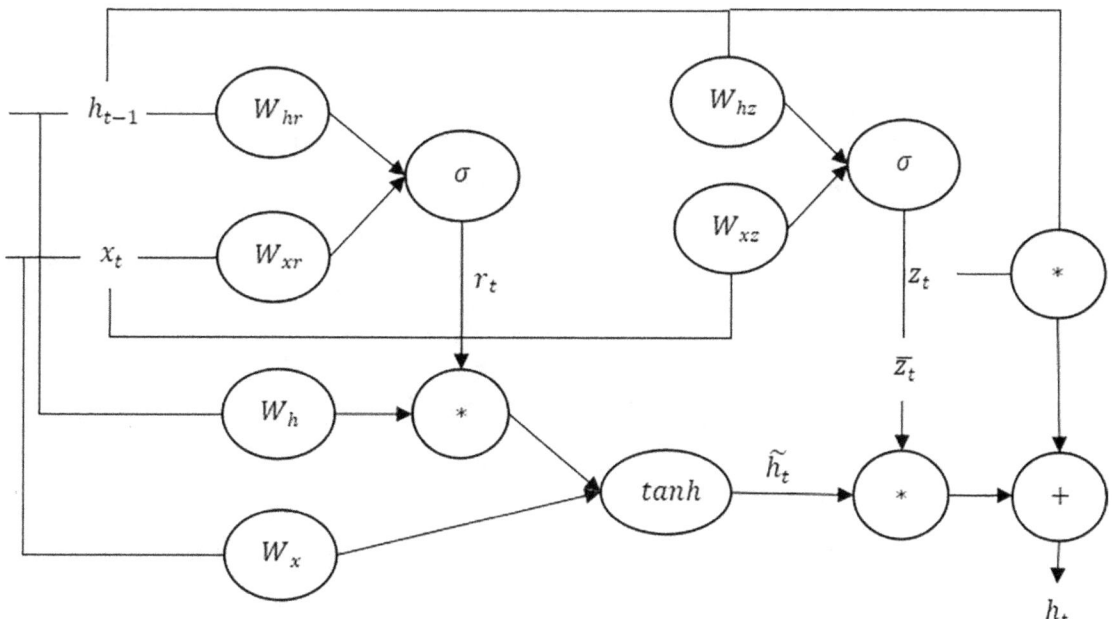

Figure 10-2. *GRU architecture*

CHAPTER 10 GATED RECURRENT UNIT AND LONG SHORT-TERM MEMORY

To summarize

- The input and h_{t-1} decide the value of the reset gate.
- The reset gate decides what portion of h_{t-1} goes into the new memory.
- The update gate depends on the input and h_{t-1}.
- The update gate decides what portion of h_{t-1} and what portion of new the memory make h_t.

Having seen the architecture of GRU, let us move to another elegant architecture called LSTM.

Long Short-Term Memory

The LSTM is a type of RNN that is capable of handling long-term dependencies. The memory cell in an LSTM can store information for a long period. The cell state is the core of LSTM and depends on the input, forget, and output gates. Let us have a brief look at the gates in an LSTM:

- Input gate (i)
- Forget gate (f)
- Output (o)

The input gate decides whether or not to write to a cell. The output gate decides how much to reveal. The forget gate tells us whether to erase a cell, and the gate tells us how much to write.

If the previous activation h_{t-1} and the input x_t are stacked, then the product of the weight W with this stacked input can become an input to various activations like sigmoid or tanh. The internal state c_t of an LSTM does not get exposed to the outside world. This c_t passed through an activation, along with o, decides the value of h_t:

$$c_t = f * c_{t-1} + i * g$$

$$h_t = o * tanh(c_t)$$

CHAPTER 10 GATED RECURRENT UNIT AND LONG SHORT-TERM MEMORY

The above process can be depicted as follows (Figure 10-3). It may be noted that people came up with their own architectures for LSTM. The architecture shown in the following figure has been adopted from https://cs231n.stanford.edu/slides/2017/cs231n_2017_lecture10.pdf [4].

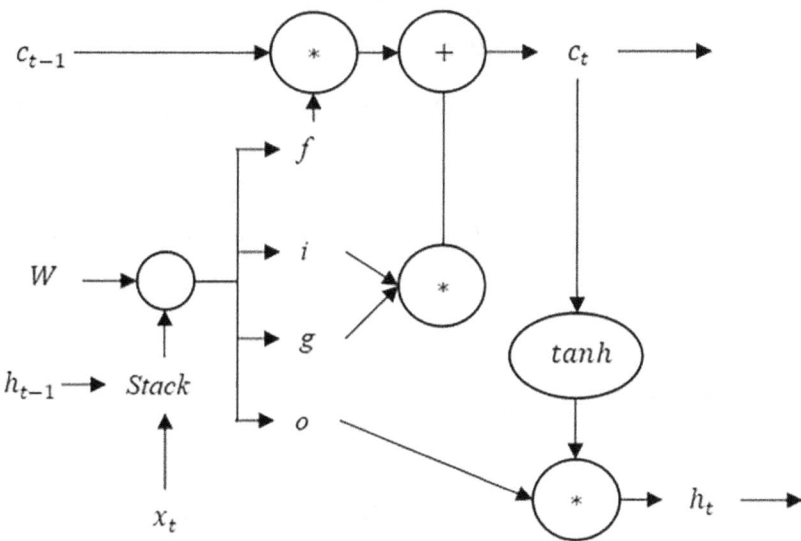

Figure 10-3. *LSTM architecture as suggested by Fei-Fei Li, Justin Johnson, and Serena Yeung [4]*

The gates of an LSTM can be used to remember the important information and forget the unnecessary ones. A brief explanation of each gate is as follows:

1. Forget Gate: The forget gate decides what information from the cell state is needed or not. It takes the previous hidden state and the current input and passes them through a sigmoid function, resulting in a value between 0 and 1 for each number in the cell state. A value of 1 indicates complete retention of the information, while a value of 0 means forgetting the information.

2. Input Gate: The input gate decides what new information should be added to the cell state. It consists of an input gate that decides which values to update and a gate that creates a vector of new candidate values that could be added to the state. The output of these two layers is combined to update the cell state.

3. Output Gate: The output gate determines what the next hidden state should be. This hidden state is used for the next time step and also for making predictions. The output gate processes the current input and the previous hidden state through a sigmoid function and then multiplies it by the tanh of the updated cell state to produce the next hidden state.

LSTMs have the capability to retain important information over long sequences and discard irrelevant data, making them very effective for tasks involving the modeling of sequential data.

Having seen the architectures of GRU and LSTM, let us now move to two important applications of these models.

Named Entity Recognition

Given a sequence of words, Named Entity Recognition (NER) aims to identify the named entities from the given sequence. It takes a sentence as an input and finds which words are the named entities. This section implements NER using the LSTM and GRU.

The CoNLL-2003 dataset is utilized in the following Listing 10-1 and contains labeled examples of sentences annotated with named entity tags. Each sentence is tokenized, and each token is tagged with an entity label. The dataset is organized in a CoNLL format, where each word in a sentence is followed by its corresponding entity tag. The sentences are separated by blank lines. Eight different models were implemented using **Keras** for NER.

These models employed different architectures of LSTM and GRU and their bidirectional variants. Each model consisted of an embedding layer, followed by recurrent layers and a dense layer with softmax activation. Each model is compiled with categorical cross-entropy loss and accuracy metrics over ten epochs. The accuracy and loss curves for each model are shown in Figure 10-4 to Figure 10-11. The code has been divided into various steps, enlisted as follows.

CHAPTER 10   GATED RECURRENT UNIT AND LONG SHORT-TERM MEMORY

Listing 10-1. Named Entity Recognition

Code:

#1. Import the CoNLL-2003 dataset from the *datasets* module using the *load_dataset* function. From *tensorflow.keras.models* import the Sequential function to create the sequential model. From *tensorflow.keras.layers* import LSTM, GRU, Bidirectional, TimeDistributed, Embedding, Dense, and Dropout layers to create models with different layers.

```
import numpy as np
from datasets import load_dataset
from tensorflow.keras.preprocessing.sequence import pad_sequences
from tensorflow.keras.utils import to_categorical
from sklearn.preprocessing import LabelEncoder
import tensorflow as tf
from tensorflow.keras.models import Sequential
from tensorflow.keras.layers import LSTM, GRU, Bidirectional, TimeDistributed, Embedding, Dense, Dropout
```

#2. Load the CoNLL-2003 dataset
```
dataset = load_dataset('conll2003', trust_remote_code=True)
```
#3. Extract the train and test data
```
train_data = dataset['train']
test_data = dataset['test']
```
#4. Create a function to extract sentences and labels from the dataset
```
def get_sentences_and_labels(data):
    sentences = [" ".join(x) for x in data['tokens']]
    labels = data['ner_tags']
    return sentences, labels
```
#5. Get sentences and labels for training and test data
```
train_sentences, train_labels = get_sentences_and_labels(train_data)
test_sentences, test_labels = get_sentences_and_labels(test_data)
```
#6. Tokenize the sentences, convert them to sequences, and pad the sequences
```
max_len = 50
```

CHAPTER 10 GATED RECURRENT UNIT AND LONG SHORT-TERM MEMORY

```
word_tokenizer = tf.keras.preprocessing.text.Tokenizer()
word_tokenizer.fit_on_texts(train_sentences)
train_sequences = word_tokenizer.texts_to_sequences(train_sentences)
test_sequences = word_tokenizer.texts_to_sequences(test_sentences)
X_train = pad_sequences(train_sequences, maxlen=max_len, padding='post')
X_test = pad_sequences(test_sequences, maxlen=max_len, padding='post')
```
#7. Encode the training and test labels
```
label_encoder = LabelEncoder()
label_encoder.fit([item for sublist in train_labels for item in sublist])
train_labels_enc = [label_encoder.transform(label) for label in
train_labels]
test_labels_enc = [label_encoder.transform(label) for label in test_labels]
```
#8. Pad the training and test labels
```
train_labels_padded = pad_sequences(train_labels_enc, maxlen=max_len,
padding='post', value=-1)
test_labels_padded = pad_sequences(test_labels_enc, maxlen=max_len,
padding='post', value=-1)
num_classes = len(label_encoder.classes_) + 1
train_labels_onehot = [to_categorical(i, num_classes=num_classes) for i in
train_labels_padded]
test_labels_onehot = [to_categorical(i, num_classes=num_classes) for i in
test_labels_padded]
y_train = np.array(train_labels_onehot)
y_test = np.array(test_labels_onehot)
```
#9. Model 1
```
model_1 = Sequential()
model_1.add(Embedding(input_dim=len(word_tokenizer.word_index) + 1, output_
dim=64, input_length=max_len))
model_1.add(GRU(units=64, return_sequences=True))
model_1.add(TimeDistributed(Dense(num_classes, activation='softmax')))
model_1.compile(optimizer='adam', loss='categorical_crossentropy',
metrics=['accuracy'])
history_1 = model_1.fit(X_train, y_train, batch_size=32, epochs=10,
validation_data=(X_test, y_test))
```
#10. Model 2

```
model_2 = Sequential()
model_2.add(Embedding(input_dim=len(word_tokenizer.word_index) + 1, output_
dim=64, input_length=max_len))
model_2.add(GRU(units=64, return_sequences=True))
model_2.add(GRU(units=64, return_sequences=True))
model_2.add(TimeDistributed(Dense(num_classes, activation='softmax')))
model_2.compile(optimizer='adam', loss='categorical_crossentropy',
metrics=['accuracy'])
history_2 = model_2.fit(X_train, y_train, batch_size=32, epochs=10,
validation_data=(X_test, y_test))
```

#11. Model 3

```
model_3 = Sequential()
model_3.add(Embedding(input_dim=len(word_tokenizer.word_index) + 1, output_
dim=64, input_length=max_len))
model_3.add(Bidirectional(GRU(units=64, return_sequences=True)))
model_3.add(TimeDistributed(Dense(num_classes, activation='softmax')))
model_3.compile(optimizer='adam', loss='categorical_crossentropy',
metrics=['accuracy'])
history_3 = model_3.fit(X_train, y_train, batch_size=32, epochs=10,
validation_data=(X_test, y_test))
```

#12. Model 4

```
model_4 = Sequential()
model_4.add(Embedding(input_dim=len(word_tokenizer.word_index) + 1, output_
dim=64, input_length=max_len))
model_4.add(Bidirectional(GRU(units=64, return_sequences=True)))
model_4.add(Bidirectional(GRU(units=64, return_sequences=True)))
model_4.add(TimeDistributed(Dense(num_classes, activation='softmax')))
model_4.compile(optimizer='adam', loss='categorical_crossentropy',
metrics=['accuracy'])
history_4 = model_4.fit(X_train, y_train, batch_size=32, epochs=10,
validation_data=(X_test, y_test))
```

#13. Model 5

```
model_5 = Sequential()
model_5.add(Embedding(input_dim=len(word_tokenizer.word_index) + 1, output_
dim=64, input_length=max_len))
```

CHAPTER 10   GATED RECURRENT UNIT AND LONG SHORT-TERM MEMORY

```
model_5.add(LSTM(units=64, return_sequences=True))
model_5.add(TimeDistributed(Dense(num_classes, activation='softmax')))
model_5.compile(optimizer='adam', loss='categorical_crossentropy',
metrics=['accuracy'])
history_5 = model_5.fit(X_train, y_train, batch_size=32, epochs=10,
validation_data=(X_test, y_test))
```

#14. Model 6

```
model_6 = Sequential()
model_6.add(Embedding(input_dim=len(word_tokenizer.word_index) + 1, output_
dim=64, input_length=max_len))
model_6.add(LSTM(units=64, return_sequences=True))
model_6.add(LSTM(units=64, return_sequences=True))
model_6.add(TimeDistributed(Dense(num_classes, activation='softmax')))
model_6.compile(optimizer='adam', loss='categorical_crossentropy',
metrics=['accuracy'])
history_6 = model_6.fit(X_train, y_train, batch_size=32, epochs=10,
validation_data=(X_test, y_test))
```

#15. Model 7

```
model_7 = Sequential()
model_7.add(Embedding(input_dim=len(word_tokenizer.word_index) + 1, output_
dim=64, input_length=max_len))
model_7.add(Bidirectional(LSTM(units=64, return_sequences=True)))
model_7.add(TimeDistributed(Dense(num_classes, activation='softmax')))
model_7.compile(optimizer='adam', loss='categorical_crossentropy',
metrics=['accuracy'])
history_7 = model_7.fit(X_train, y_train, batch_size=32, epochs=10,
validation_data=(X_test, y_test))
```

#16. Model 8

```
model_8 = Sequential()
model_8.add(Embedding(input_dim=len(word_tokenizer.word_index) + 1, output_
dim=64, input_length=max_len))
model_8.add(Bidirectional(LSTM(units=64, return_sequences=True)))
model_8.add(Bidirectional(LSTM(units=64, return_sequences=True)))
model_8.add(TimeDistributed(Dense(num_classes, activation='softmax')))
```

```python
model_8.compile(optimizer='adam', loss='categorical_crossentropy', metrics=['accuracy'])
history_8 = model_8.fit(X_train, y_train, batch_size=32, epochs=10, validation_data=(X_test, y_test))
```

#17. Create a function to plot the training and validation loss and accuracy

```python
def plot_history(history, model_name):
    plt.figure(figsize=(12, 6))
    plt.subplot(1, 2, 1)
    plt.plot(history.history['accuracy'])
    plt.plot(history.history['val_accuracy'])
    plt.title(f'{model_name} Model Accuracy')
    plt.xlabel('Epoch')
    plt.ylabel('Accuracy')
    plt.legend(['Train', 'Val'], loc='upper left')
    plt.subplot(1, 2, 2)
    plt.plot(history.history['loss'])
    plt.plot(history.history['val_loss'])
    plt.title(f'{model_name} Model Loss')
    plt.xlabel('Epoch')
    plt.ylabel('Loss')
    plt.legend(['Train', 'Val'], loc='upper left')
    plt.tight_layout()
    plt.show()
```

#18. Plot the accuracy and loss curves for each model

```python
plot_history(history_1, "Model 1")
plot_history(history_2, "Model 2")
plot_history(history_3, "Model 3")
plot_history(history_4, "Model 4")
plot_history(history_5, "Model 5")
plot_history(history_6, "Model 6")
plot_history(history_7, "Model 7")
plot_history(history_8, "Model 8")
```

#19. Create a function to calculate mean validation accuracy

```python
def mean_validation_accuracy(history):
```

```
    val_acc = history.history['val_accuracy']
    mean_acc = np.mean(val_acc)
    return mean_acc
#20. Calculate the mean validation accuracy for each model
mean_acc_1 = mean_validation_accuracy(history_1)
mean_acc_2 = mean_validation_accuracy(history_2)
mean_acc_3 = mean_validation_accuracy(history_3)
mean_acc_4 = mean_validation_accuracy(history_4)
mean_acc_5 = mean_validation_accuracy(history_5)
mean_acc_6 = mean_validation_accuracy(history_6)
mean_acc_7 = mean_validation_accuracy(history_7)
mean_acc_8 = mean_validation_accuracy(history_8)
```

Output:

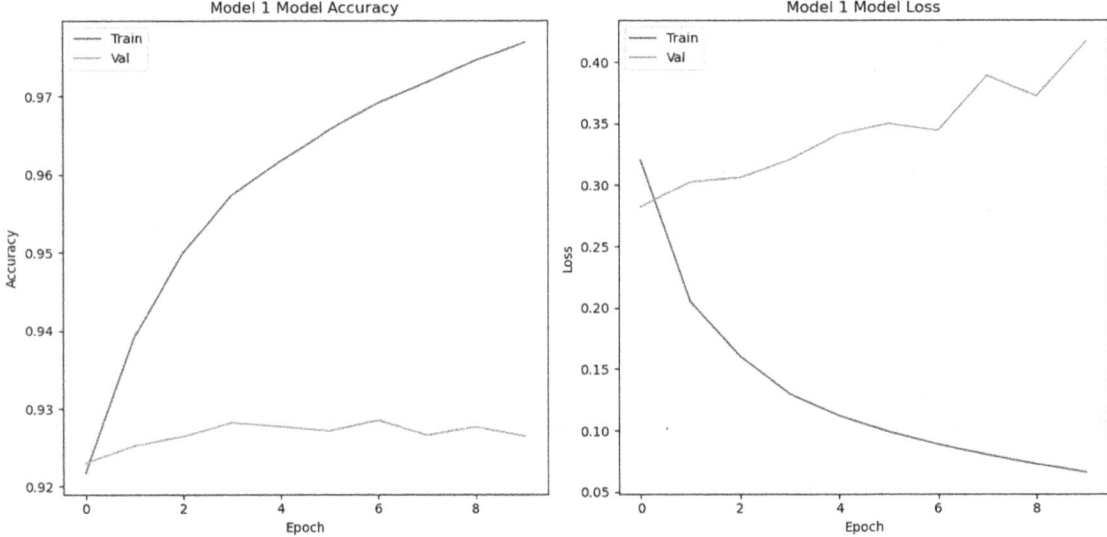

Figure 10-4. *Loss and accuracy curves: Model 1*

CHAPTER 10 GATED RECURRENT UNIT AND LONG SHORT-TERM MEMORY

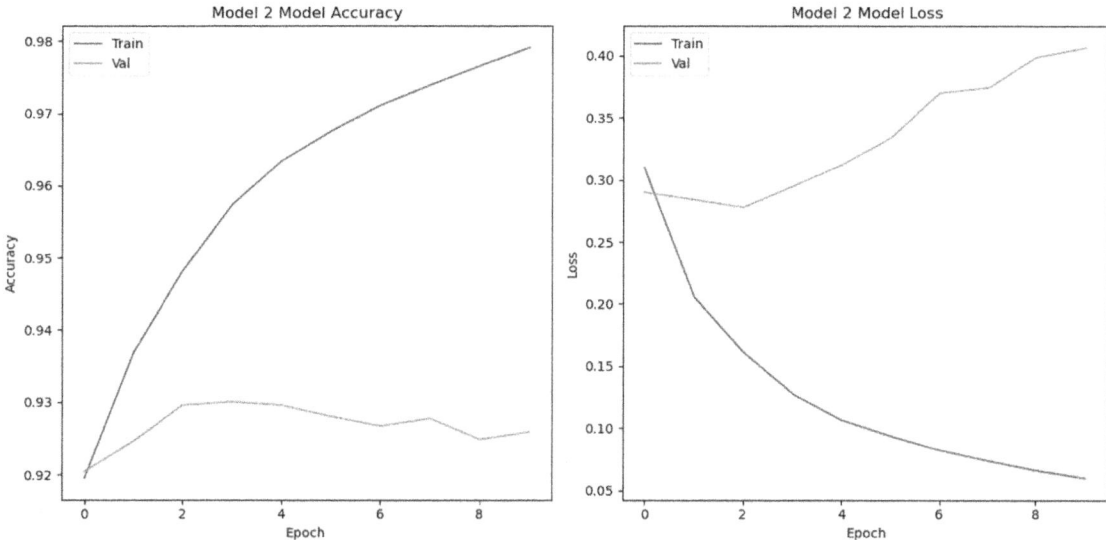

Figure 10-5. *Loss and accuracy curves: Model 2*

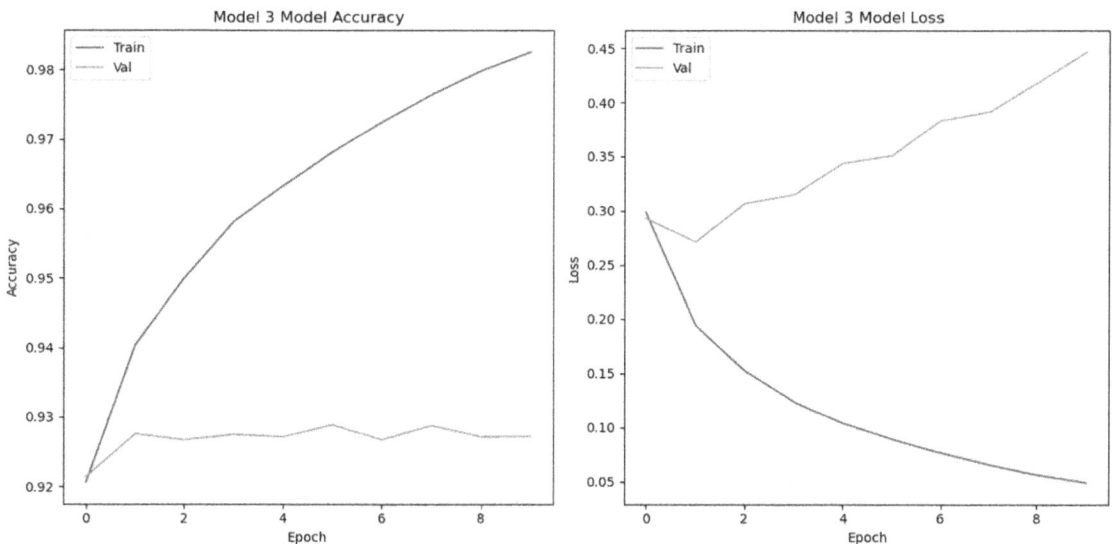

Figure 10-6. *Loss and accuracy curves: Model 3*

CHAPTER 10 GATED RECURRENT UNIT AND LONG SHORT-TERM MEMORY

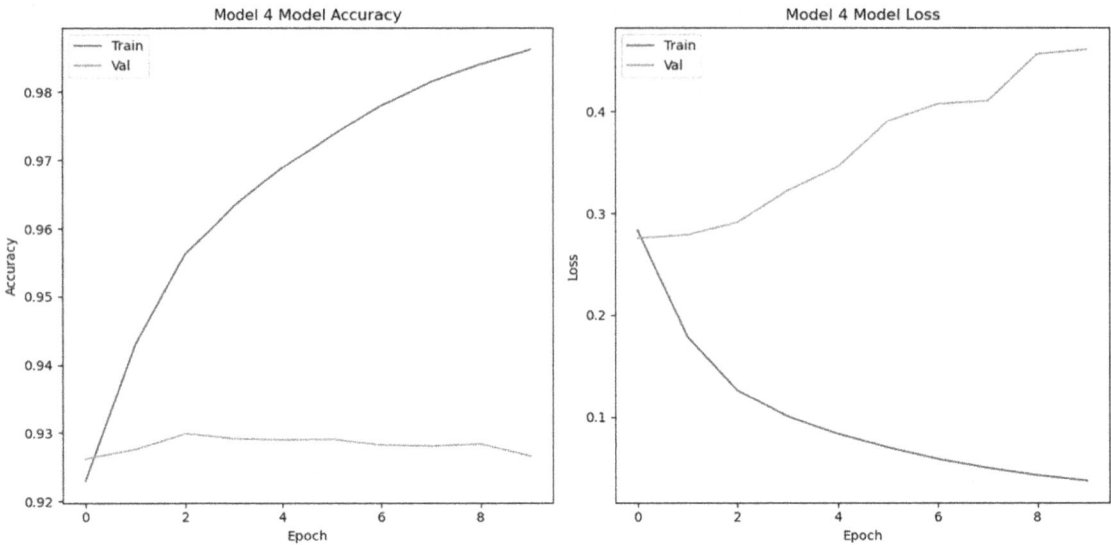

Figure 10-7. *Loss and accuracy curves: Model 4*

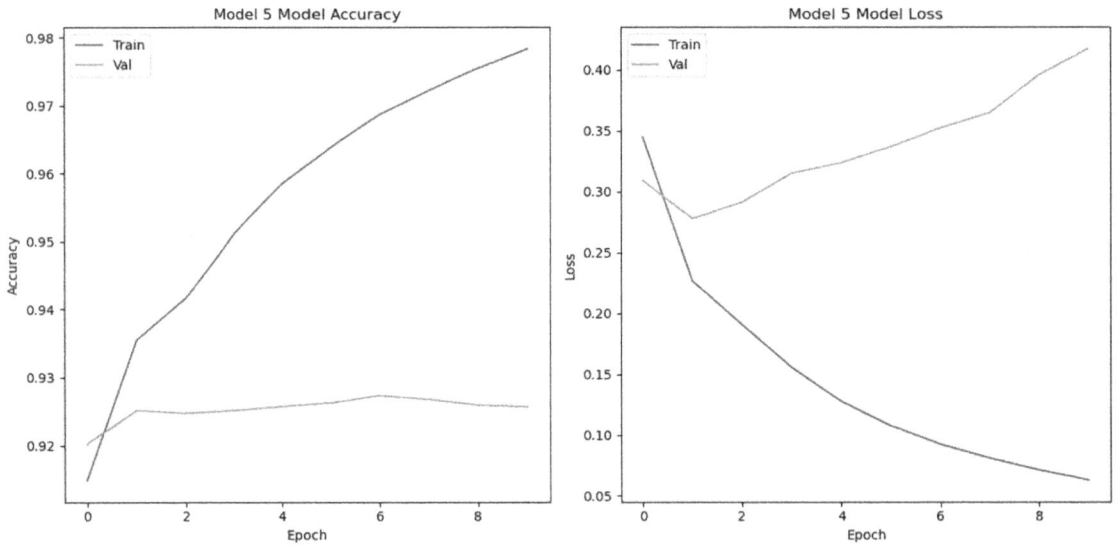

Figure 10-8. *Loss and accuracy curves: Model 5*

CHAPTER 10 GATED RECURRENT UNIT AND LONG SHORT-TERM MEMORY

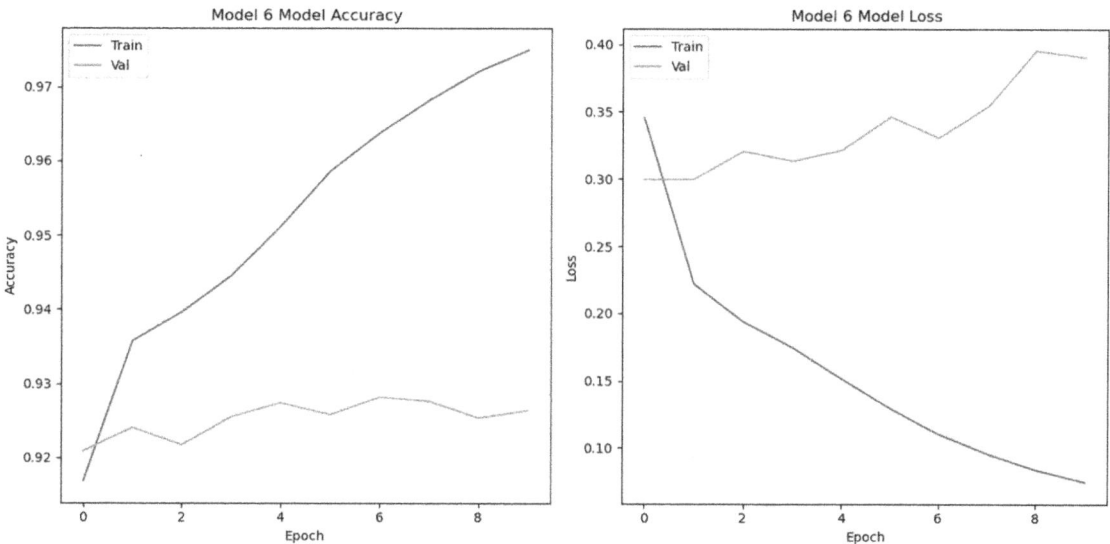

Figure 10-9. *Loss and accuracy curves: Model 6*

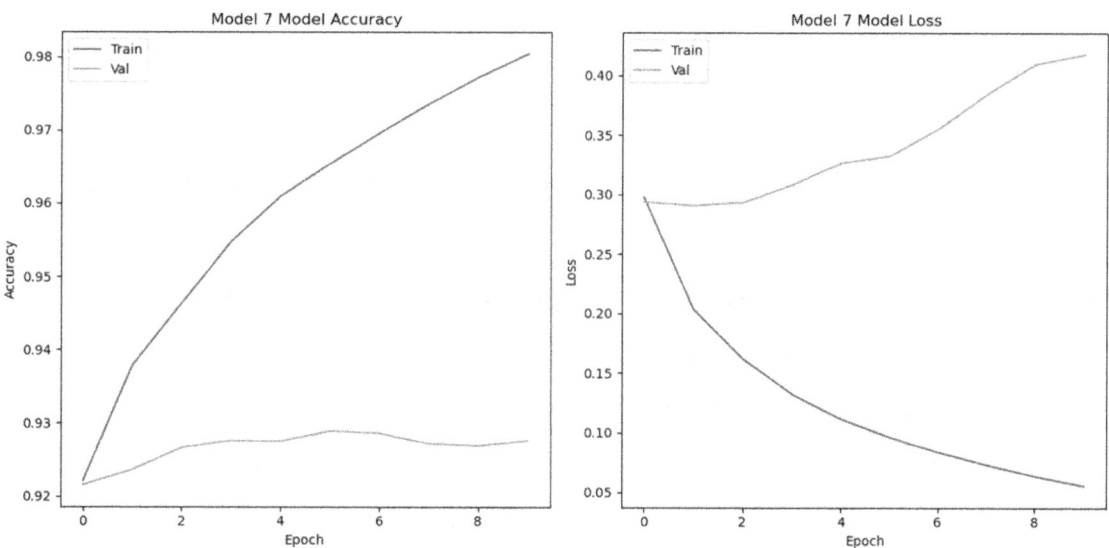

Figure 10-10. *Loss and accuracy curves: Model 7*

CHAPTER 10 GATED RECURRENT UNIT AND LONG SHORT-TERM MEMORY

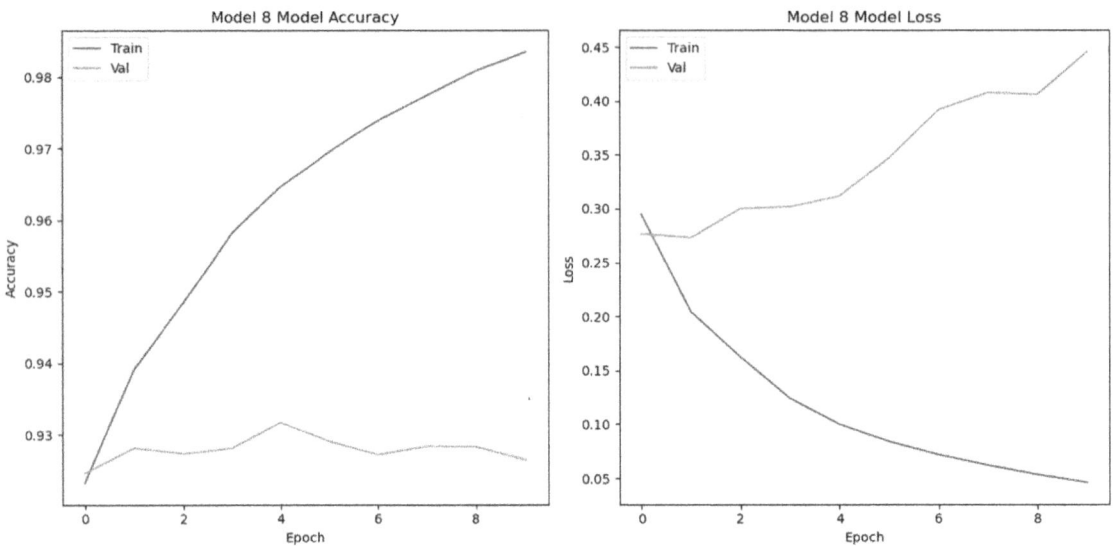

Figure 10-11. *Loss and accuracy curves: Model 8*

The results of the above experiments are summarized in Table 10-1.

Table 10-1. *Mean Validation Accuracy of Eight Different Models*

Architecture	Mean Validation Accuracy
GRU with a single layer having 64 units	0.9267
Stacked GRU with two layers having 64 units in each layer	0.9268
Bidirectional GRU with a single layer having 64 units	0.9269
Stacked bidirectional GRU with two layers having 64 units in each layer	0.9282
LSTM with a single layer having 64 units	0.9253
Stacked LSTM with two layers having 64 units in each layer	0.9253
Bidirectional LSTM with a single layer having 64 units	0.9266
Stacked Bidirectional LSTM with two layers having 64 units in each layer	0.9279

Let us have a look at the use of LSTM and GRU models in sentiment classification.

CHAPTER 10　GATED RECURRENT UNIT AND LONG SHORT-TERM MEMORY

Sentiment Classification

Refer to Listing 9-1 of the previous chapter that classifies the given sentences into positive or negative sentiments. The four different RNN models were created and evaluated on the IMDB movie review sentiment dataset. This dataset contains 50,000 movie reviews, evenly split into positive and negative sentiments. The dataset was preprocessed by removing stop words, tokenizing, and padding the reviews. The following Listing 10-2 implements eight different models using **Keras** for sentiment classification. These models employed different architectures of LSTM and GRU and their bidirectional variants. Each model consisted of an embedding layer, followed by recurrent layers and a dense layer with sigmoid activation. The variation of accuracy and loss with the number of epochs is shown in figures from Figure 10-12 to Figure 10-19.

Listing 10-2. Sentiment classification using the IMDB dataset

Code:

#1. Import the IMDB dataset from *tensorflow.keras.datasets*, stopwords from *nltk.corpus*. The *tensorflow.keras.models, tensorflow.keras.layers* are imported to design a sequential model having Embedding, GRU, LSTM, and Bidirectional layers.

```
import numpy as np
from tensorflow.keras.datasetsimportimdb
from tensorflow.keras.preprocessing.sequence import pad_sequences
from tensorflow.keras.preprocessing.text import Tokenizer
from nltk.corpus import stopwords
import nltk
from tensorflow.keras.models import Sequential
from tensorflow.keras.layers import Embedding, GRU, Dense, Bidirectional, LSTM
from matplotlib import pyplot as plt
```

#2. The stopwords are downloaded from NLTK

```
nltk.download('stopwords')
```

#3. The IMDB dataset is downloaded and limited to the top 10,000 most frequent words.

```
max_features = 10000
```

CHAPTER 10　GATED RECURRENT UNIT AND LONG SHORT-TERM MEMORY

```
(X_train, y_train), (X_test, y_test) = imdb.load_data(num_words=max_
features)
```
#4. Create a reverse dictionary to decode reviews back to words
```
word_index = imdb.get_word_index()
reverse_word_index = dict([(value, key) for (key, value) in word_index.
items()])
```
#5. Create a function to decode reviews from sequences of integers to words
```
def decode_review(encoded_review):
    return ' '.join([reverse_word_index.get(i - 3, '?') for i in encoded_
review])
```
#6. Decode all reviews in the training and test sets
```
decoded_train = [decode_review(review) for review in X_train]
decoded_test = [decode_review(review) for review in X_test]
```
#7. Remove the stop words from the reviews
```
stop_words = set(stopwords.words('english'))
def remove_stop_words(text):
    return ' '.join([word for word in text.split() if word not in
stop_words])
cleaned_train = [remove_stop_words(review) for review in decoded_train]
cleaned_test = [remove_stop_words(review) for review in decoded_test]
```
#8. Tokenize the cleaned reviews using the Tokenizer function imported from *tensorflow.keras.preprocessing.text*
```
tokenizer = Tokenizer(num_words=max_features)
tokenizer.fit_on_texts(cleaned_train)
```
#9. Convert the tokenized reviews to sequences
```
train_sequences = tokenizer.texts_to_sequences(cleaned_train)
test_sequences = tokenizer.texts_to_sequences(cleaned_test)
```
#10. Pad the sequences to ensure they all have the same length
```
maxlen = 100
X_train = pad_sequences(train_sequences, maxlen=maxlen)
X_test = pad_sequences(test_sequences, maxlen=maxlen)
```
#11. Create a function to create, compile, and train a model
```
def compile_and_train(model, epochs=10):
    model.compile(optimizer='adam', loss='binary_crossentropy',
metrics=['acc'])
```

```
    history = model.fit(X_train, y_train, epochs=epochs, batch_
size=32,validation_split=0.3)
    return history
```
#12. Model 1
```
model_1 = Sequential()
model_1.add(Embedding(max_features, 32))
model_1.add(GRU(32))
model_1.add(Dense(1, activation='sigmoid'))
history_1 = compile_and_train(model_1)
```
#13. Model 2
```
model_2 = Sequential()
model_2.add(Embedding(max_features, 32))
model_2.add(GRU(32, return_sequences=True))
model_2.add(GRU(32))
model_2.add(Dense(1, activation='sigmoid'))
history_2 = compile_and_train(model_2)
```
#14. Model 3
```
model_3 = Sequential()
model_3.add(Embedding(max_features, 32))
model_3.add(Bidirectional(GRU(32)))
model_3.add(Dense(1, activation='sigmoid'))
history_3 = compile_and_train(model_3)
```
#15. Model 4
```
model_4 = Sequential()
model_4.add(Embedding(max_features, 32))
model_4.add(Bidirectional(GRU(32, return_sequences=True)))
model_4.add(Bidirectional(GRU(32)))
model_4.add(Dense(1, activation='sigmoid'))
history_4 = compile_and_train(model_4)
```
#16. Model 5
```
model_5 = Sequential()
model_5.add(Embedding(max_features, 32))
model_5.add(LSTM(32))
model_5.add(Dense(1, activation='sigmoid'))
history_5 = compile_and_train(model_5)
```

#17. Model 6
```
model_6 = Sequential()
model_6.add(Embedding(max_features, 32))
model_6.add(LSTM(32, return_sequences=True))
model_6.add(LSTM(32))
model_6.add(Dense(1, activation='sigmoid'))
history_6 = compile_and_train(model_6)
```
#18. Model 7
```
model_7 = Sequential()
model_7.add(Embedding(max_features, 32))
model_7.add(Bidirectional(LSTM(32)))
model_7.add(Dense(1, activation='sigmoid'))
history_7 = compile_and_train(model_7)
```
#19. Model 8
```
model_8 = Sequential()
model_8.add(Embedding(max_features, 32))
model_8.add(Bidirectional(LSTM(32, return_sequences=True)))
model_8.add(Bidirectional(LSTM(32)))
model_8.add(Dense(1, activation='sigmoid'))
history_8 = compile_and_train(model_8)
```
#20. Create a function to plot the training and validation loss and accuracy
```
def plot_history(history, model_name):
    plt.figure(figsize=(12, 6))
    plt.subplot(1, 2, 1)
    plt.plot(history.history['accuracy'])
    plt.plot(history.history['val_accuracy'])
    plt.title(f'{model_name} Model Accuracy')
    plt.xlabel('Epoch')
    plt.ylabel('Accuracy')
    plt.legend(['Train', 'Val'], loc='upper left')
    plt.subplot(1, 2, 2)
    plt.plot(history.history['loss'])
    plt.plot(history.history['val_loss'])
    plt.title(f'{model_name} Model Loss')
```

```
    plt.xlabel('Epoch')
    plt.ylabel('Loss')
    plt.legend(['Train', 'Val'], loc='upper left')
    plt.tight_layout()
    plt.show()
#21. Plot the accuracy and loss curves for each model
plot_history(history_1, "Model 1")
plot_history(history_2, "Model 2")
plot_history(history_3, "Model 3")
plot_history(history_4, "Model 4")
plot_history(history_5, "Model 5")
plot_history(history_6, "Model 6")
plot_history(history_7, "Model 7")
plot_history(history_8, "Model 8")
#22. Create a function to calculate mean validation accuracy
def mean_validation_accuracy(history):
    val_acc = history.history['val_accuracy']
    mean_acc = np.mean(val_acc)
    return mean_acc
#23. Calculate the mean validation accuracy for each model
mean_acc_1 = mean_validation_accuracy(history_1)
mean_acc_2 = mean_validation_accuracy(history_2)
mean_acc_3 = mean_validation_accuracy(history_3)
mean_acc_4 = mean_validation_accuracy(history_4)
mean_acc_5 = mean_validation_accuracy(history_5)
mean_acc_6 = mean_validation_accuracy(history_6)
mean_acc_7 = mean_validation_accuracy(history_7)
mean_acc_8 = mean_validation_accuracy(history_8)
```

Output:

CHAPTER 10 GATED RECURRENT UNIT AND LONG SHORT-TERM MEMORY

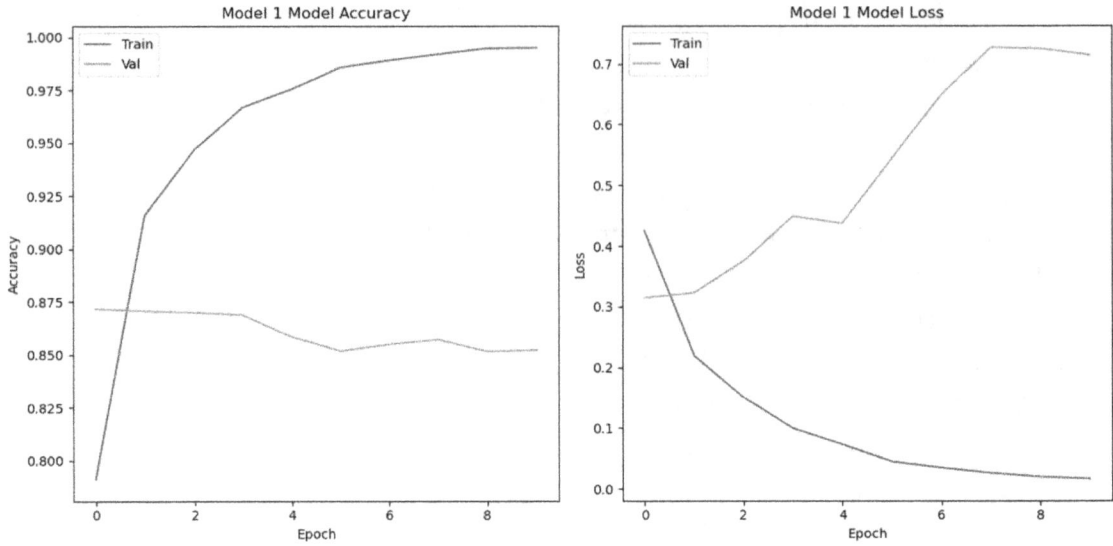

Figure 10-12. *Loss and accuracy curves: Model 1*

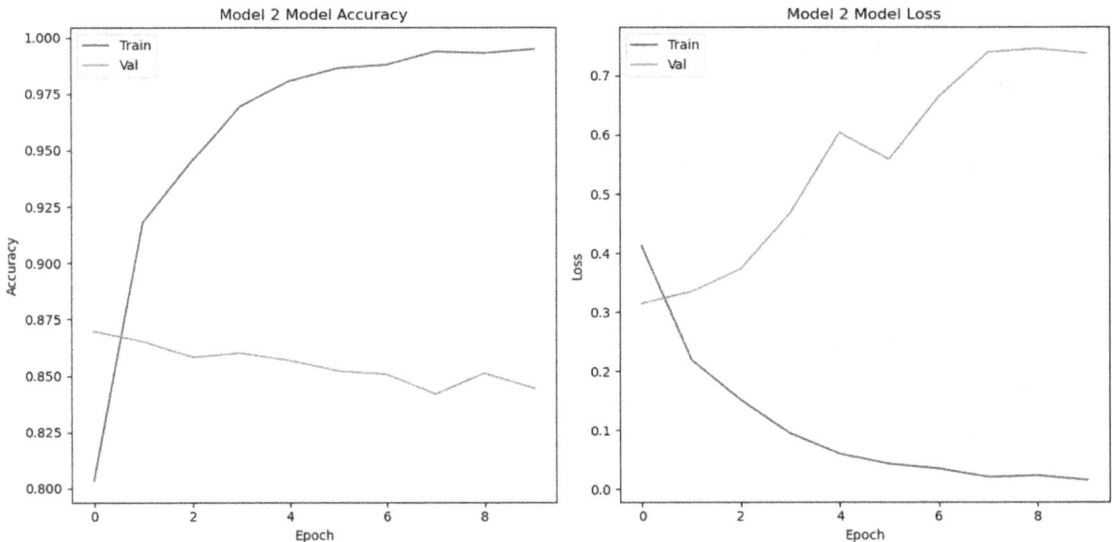

Figure 10-13. *Loss and accuracy curves: Model 2*

CHAPTER 10　GATED RECURRENT UNIT AND LONG SHORT-TERM MEMORY

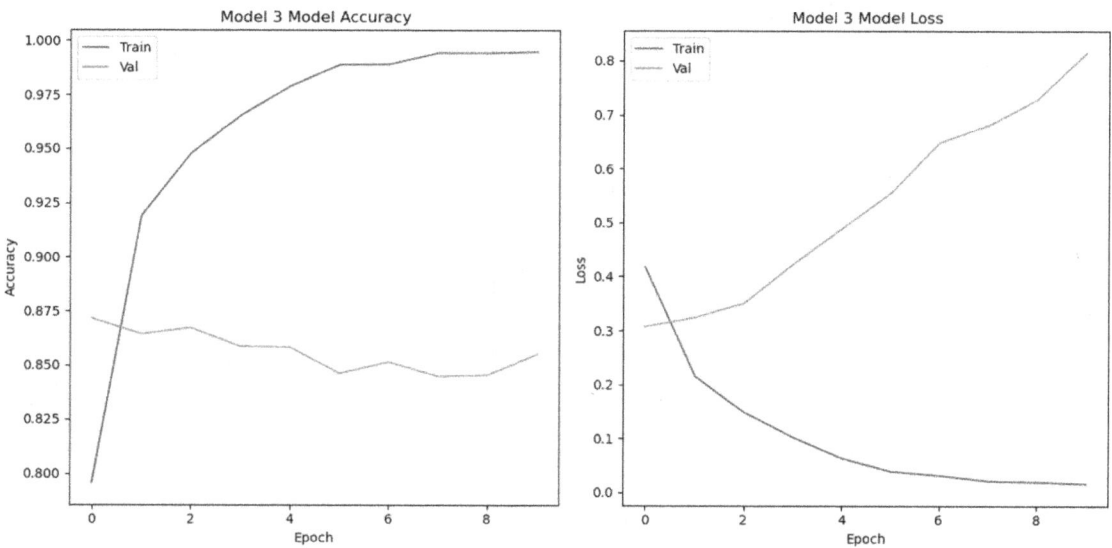

Figure 10-14. *Loss and accuracy curves: Model 3*

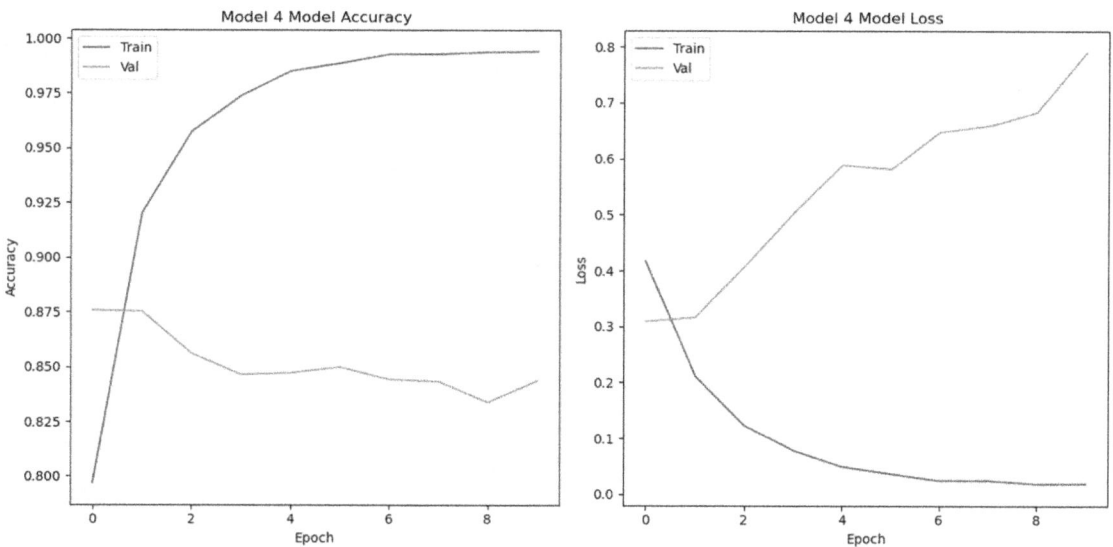

Figure 10-15. *Loss and accuracy curves: Model 4*

CHAPTER 10　GATED RECURRENT UNIT AND LONG SHORT-TERM MEMORY

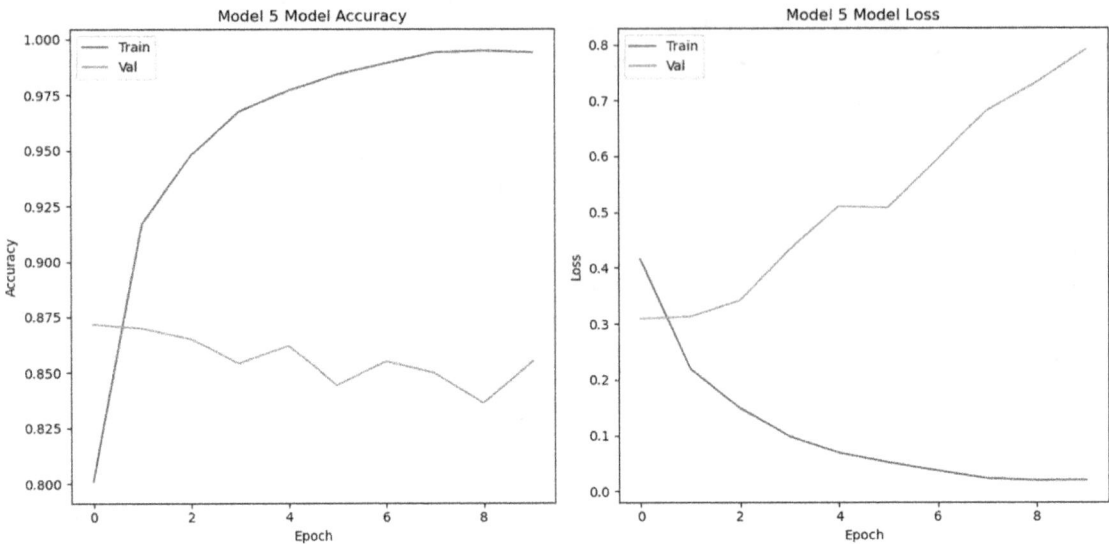

Figure 10-16. *Loss and accuracy curves: Model 5*

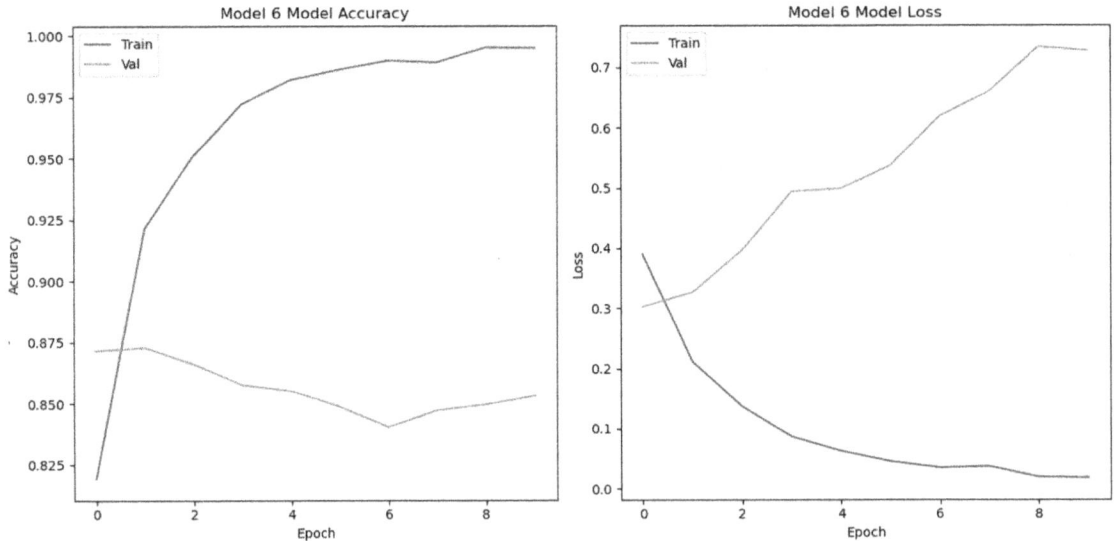

Figure 10-17. *Loss and accuracy curves: Model 6*

CHAPTER 10 GATED RECURRENT UNIT AND LONG SHORT-TERM MEMORY

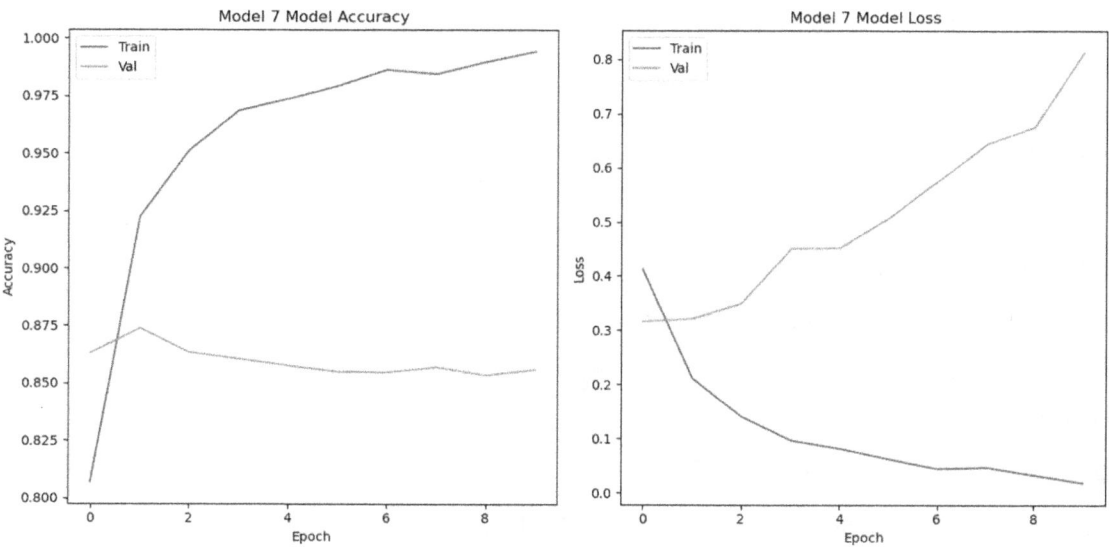

Figure 10-18. *Loss and accuracy curves: Model 7*

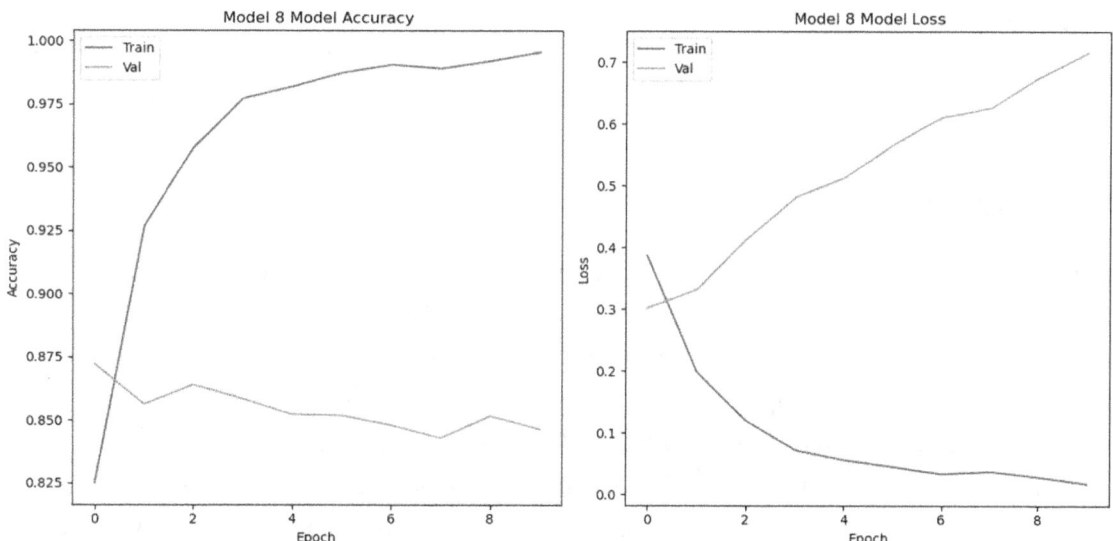

Figure 10-19. *Loss and accuracy curves: Model 8*

The results of the above experiments are summarized in Table 10-2.

Table 10-2. *Mean Validation Accuracy of Eight Different Models*

Architecture	Mean Validation Accuracy
GRU with a single layer having 64 units	0.8607
Stacked GRU with two layers having 64 units in each layer	0.8549
Bidirectional GRU with a single layer having 64 units	0.8563
Stacked bidirectional GRU with two layers having 64 units in each layer	0.8516
LSTM with a single layer having 64 units	0.8563
Stacked LSTM with two layers having 64 units in each layer	0.8562
Bidirectional LSTM with a single layer having 64 units	0.8594
Stacked bidirectional LSTM with two layers having 64 units in each layer	0.8544

Conclusion

The chapter begins by discussing the necessity for models such as Long Short-Term Memory and Gated Recurrent Units. It highlights their importance in handling the limitations of traditional Recurrent Neural Networks. An informed discussion on LSTM and GRU follows, explaining their architecture and functionalities. The chapter then explores the application of these models in Named Entity Recognition and Sentiment Analysis, demonstrating their effectiveness in processing and understanding text data. Additionally, fascinating applications of LSTM and GRU are presented in the Appendix C and Appendix D of this book, providing practical insights into their usage. The attention models introduced in the last chapter set the basis of the transformer models. These models are the present and the future of models that can efficiently and effectively deal with sequences. Readers are encouraged to engage with the exercises at the end of the chapter to reinforce their understanding and gain hands-on experience with these concepts.

Exercises
Multiple-Choice Questions

1. Which of the following are present in LSTM?

    a. Forget gate

    b. Update gate

    c. Both

    d. None

2. What is the difference between GRU and LSTM?

    a. GRU has a forget gate; LSTM does not.

    b. LSTM has a forget gate; GRU does not.

    c. GRU and LSTM are identical.

    d. LSTM has no gates.

3. Which of the following can handle the vanishing gradient problem?

    a. LSTM

    b. GRU

    c. Both

    d. None

4. LSTM works better in

    a. Image classification

    b. Sentiment classification

    c. Regression

    d. Clustering

CHAPTER 10   GATED RECURRENT UNIT AND LONG SHORT-TERM MEMORY

5. Which of the following is a sequence model?

   a. RNN

   b. LSTM

   c. GRU

   d. All of the above

6. LSTM uses which of the following activation functions?

   a. ReLU

   b. Sigmoid

   c. Tanh

   d. Both b and c

7. How is the hidden state updated in LSTM?

   a. Using only the input gate

   b. Using the output gate and forget gate

   c. Using the input gate, forget gate, and output gate

   d. Using a single gate

8. Which algorithm is used for training a sequence model?

   a. Gradient descent

   b. Backpropagation Through Time (BPTT)

   c. Stochastic gradient descent (SGD)

   d. Reinforcement Learning

9. Image captioning can be done using

   a. Neural Network

   b. Sequence model

   c. Bag of Words

   d. All of the above

10. Which of LSTM and GRU is better?

    a. LSTM

    b. GRU

    c. Both are equally good

    d. Depends on the specific task

Theory

1. Explain the problems in RNN. How can these problems be handled using a GRU?

2. Explain the architecture of GRU. What is the difference between GRU and LSTM?

3. Explain the architecture of LSTM. What is the significance of each gate?

4. How is a GRU different from an LSTM? State which can be used in which case?

Application-Based Questions

1. Write a program to generate text using a character-based Recurrent Neural Network (RNN). You will use a dataset of Shakespeare's writing from Andrej Karpathy's article, "The Unreasonable Effectiveness of Recurrent Neural Networks." The goal is to train a model that can predict the next character in a sequence of characters from this data. You can use the trained model to generate longer sequences of text by predicting one character at a time. The dataset can be found at this link: Kaggle Shakespeare Text Generation with RNN.

2. In the above question, analyze the effect of the following on the performance of the model:

   a. Number of layers

   b. Number of units in the embedding layer

   c. Use of RNN, Bi-RNN, GRU, and LSTM

   d. Optimizers

3. Now develop a next word generation model (instead of the next character generation model) and assess if the model performs better now.

References

[1] Manning, C., Socher, R., Mohammadi, M., Mundra, R., Wang, L. & Kamath, A. (2019). *CS224n: Natural Language Processing with Deep Learning.* https://web.stanford.edu/class/cs224n/readings/cs224n-2019-notes05-LM_RNN.pdf

[2] See, A., Hewitt, J., Manning, C. & Socher, R. Natural Language Processing with Deep Learning [Lecture]. In *CS224N/Ling284* (2013). https://web.stanford.edu/class/archive/cs/cs224n/cs224n.1204/slides/cs224n-2020-lecture07-fancy-rnn.pdf

[3] Arnold, T. B. (2016). *Recurrent neural networks.* https://euler.stat.yale.edu/~tba3/stat665/lectures/lec21/lecture21.pdf

[4] Stanford. (n.d.). *Lecture 10: Recurrent neural networks.* https://cs231n.stanford.edu/slides/2017/cs231n_2017_lecture10.pdf

CHAPTER 11

Autoencoders

Introduction

An autoencoder may be considered as a network that implements the identity function. It takes a data sample as the input and also considers the same sample as the output, thus behaving as a self-supervised model. Though the idea of a network aiming to replicate its input seems pointless, it may be used for many purposes such as

- Creating embeddings of the input data such as images and audio data
- Data compression
- Generating new data similar to the input data and so on

This chapter explains the basics of autoencoders and then moves to the implementation of a basic autoencoder that replicates the MNIST dataset. This replication requires due deliberation and careful selection of hyperparameters, as demonstrated in the next program, which accomplishes the same task using the CIFAR-10 dataset. We then move to different variants of autoencoders and discuss sparse, denoising, and variational autoencoders. The last section concludes.

Concept and Types

Consider a network having many hidden layers between the input and the output layer. These layers are arranged in a way that the number of neurons in the second layer is the same as the second last layer; the number of neurons in the third layer is the same as the number of neurons in the third last layer; and so on. Also, assuming that the number of layers in such a network is odd, then the middle layer may be used to extract the latent representation of the input and hence be considered the most important one.

Chapter 11 Autoencoders

The autoencoder consists of two sub-networks: encoder and decoder. The encoder converts the input data into a compact representation, and the decoder converts this compact representation again to the output, which is the same as the input.

The Math

Consider a network that can act as an identity function but regenerates the input via an intermediate layer. That is, if the input is x, then the output of the intermediate layer will be

$$h = f(W_x x + b_x)$$

where f is an activation function and W_x and b_x are the weights and the bias, respectively. The output of the intermediate layer is then multiplied by the weight W_h and passes through an activation function g to produce $\hat{x}$:

$$\hat{x} = g(W_h h + b_h)$$

The aim of the network is to reduce the difference between x and $\hat{x}$, that is, to minimize the loss. $L = \frac{1}{2}(\hat{x} - x)^2$, in the case if the inputs are real.

If the inputs are binary, then the loss is taken as

$$L = -\sum_i x_i \log \hat{x}_i + (1 - x_i) \log (1 - \hat{x}_i)$$

So the network encodes the input x to a latent representation h and decodes h back to $\hat{x}$ with the aim of making $\hat{x}$ the same as x. The model is trained to minimize loss with respect to parameters W_x, b_x, W_h, and b_h.

Types of Autoencoders

Autoencoders may be segregated into two types: under-complete and over-complete, based on the size of the hidden layer.

Under-complete Autoencoder

Under-complete autoencoders have the number of units in the hidden layer fewer than that in the input layer. An example of such an autoencoder is shown in Figure 11-1. After training, if the input x_i can be reconstructed exactly by the network, it implies that the embedding contains a good enough latent representation of the input. This is referred to as lossless embedding.

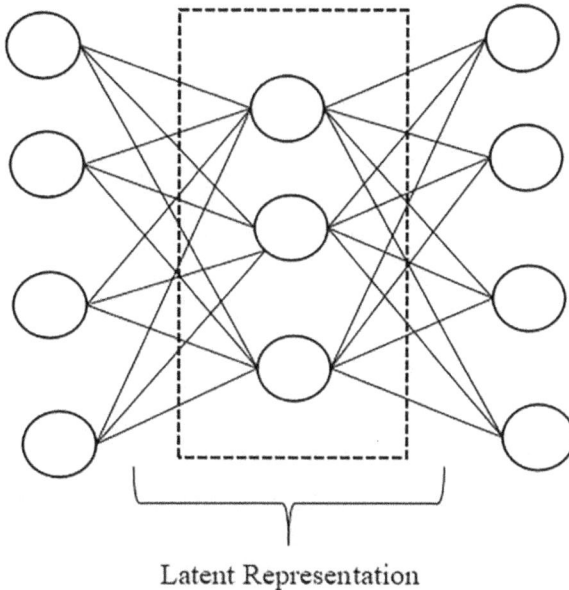

Figure 11-1. Under-complete autoencoder

Over-complete Autoencoder

An over-complete autoencoder has the number of units in the hidden layer of the encoder part more than that in the input layer. An example of such a network is shown in Figure 11-2. This type of encoder generally performs regularization and incorporates sparsity. In over-complete autoencoders, it is possible that we simply copy the values of x in the first few cells of the hidden layer and then use them for reconstruction. The over-complete autoencoder has to ensure that this does not happen.

CHAPTER 11   AUTOENCODERS

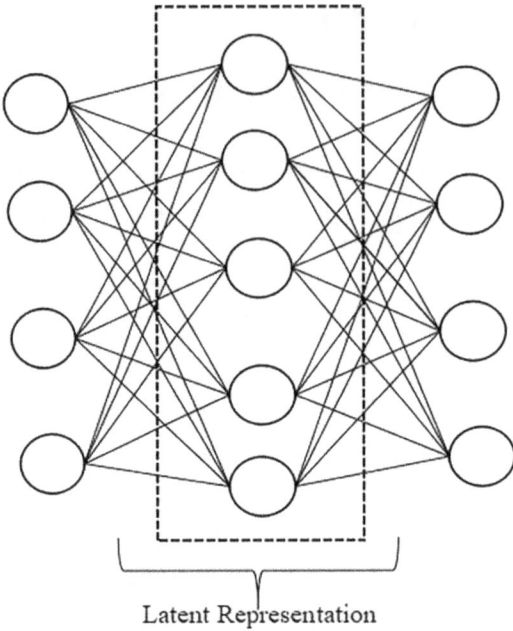

Figure 11-2. Over-complete autoencoder

These autoencoders can perform data compression like Principal Component Analysis. Let us explore the similarities and differences between the two methods of data compression.

Autoencoder and Principal Component Analysis

Principal Component Analysis (PCA) transforms the original data having various features into a new set of features called principal components. They capture the maximum variance in the data within fewer dimensions or features. This is particularly useful for dimensionality reduction. We can find the PCA of the given data using the following method:

1. For the input data X, we find the mean deviation ($X - \overline{X}$), followed by the formation of a scatter matrix by multiplying the mean deviation with its transpose:

$$\text{Scatter Matrix} = (X - \overline{X})(X - \overline{X})^T$$

2. The eigenvalues of the scatter matrix so formed are then obtained.

290

3. The first "d" eigenvalues in decreasing order are then found and corresponding vectors are concatenated.

4. The transformation matrix so formed is then multiplied with the original matrix so as to obtain the transformed features.

Note that the transformed features are such that the first feature captures maximum variance and subsequent features capture the remaining variance. As stated earlier, both PCA and autoencoders can be used for dimensionality reduction; however, there are notable differences between the two.

PCA is a linear dimensionality reduction method. As stated earlier, it finds the direction in which the variance is maximum. However, such methods do not work well for the datasets wherein the relationship between the variables is not linear. In addition to the above, the principal components of PCA are easy to interpret. An autoencoder, on the other hand, is a nonlinear data reduction technique. This makes it powerful as it can even work for datasets in which the relation between the variables is nonlinear. However, it is difficult to interpret the latent representations created by autoencoders. The computational complexity of autoencoders is more as compared with PCA, but they can generate a very good reconstruction of the original data.

Training of an Autoencoder

In training an autoencoder we take the first hidden layer of the encoder part along with the input, which is the same as the output, and create a new network as shown in Figure 11-3(a).

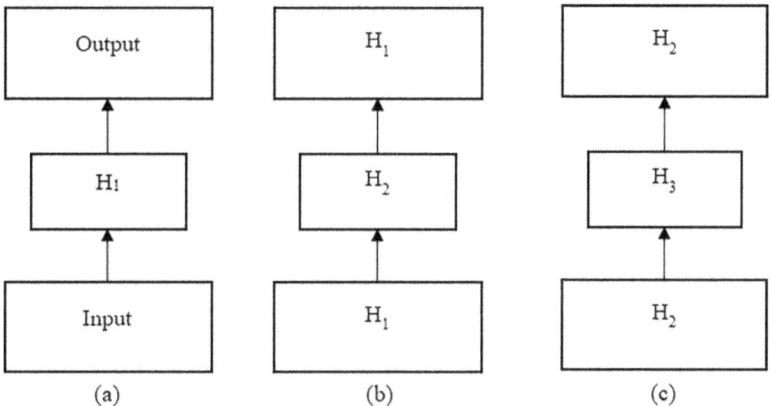

Figure 11-3. *Learning the weights of the first, second, and third hidden layers*

CHAPTER 11 AUTOENCODERS

After training this network, we obtain the weights of H_1. Once the weights of H_1 have been obtained, we take H_1 as the input and the output of the new network and learn H_2 as shown in Figure 11-3(b).

Likewise, we can learn the weights of H_3 (Figure 11-3(c)) and so on and then construct the whole network (Figure 11-4).

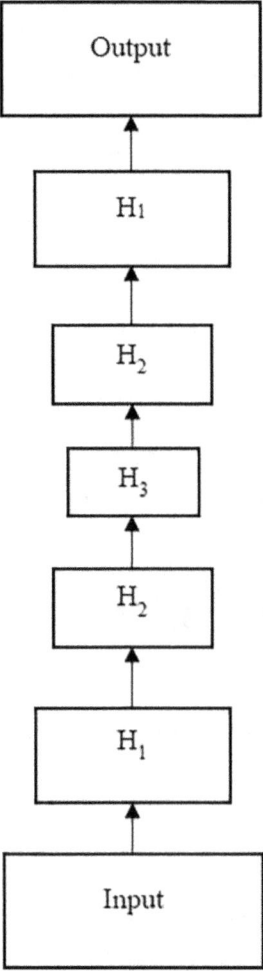

Figure 11-4. *An autoencoder containing three hidden layers*

The weights of H_1, H_2, and H_3 are learned separately. Let us now have a look at some interesting applications of autoencoders.

Latent Representation Using Autoencoders
Experiment 1

As stated earlier, an autoencoder can be used to find an effective encoding for a given input. We generally use an under-complete autoencoder for this purpose. As an example, the following program reconstructs the MNIST dataset using an autoencoder in (Listing 11-1). To accomplish the task, we carry out the following steps.

Listing 11-1. Reconstructing the MNIST dataset using an autoencoder

Code:
We import the requisite libraries to a) create the model b) Plot the performance and loss curves c) Plot the images d) Carry out low-level numeric tasks.
```
import tensorflow as tf
from tensorflow.keras import datasets, layers, models
import matplotlib.pyplot as plt
import numpy as np
from tensorflow.keras import optimizers
```
We then split the dataset into train and test.
```
mnist_data=tf.keras.datasets.mnist
(X_train,y_train),(X_test,y_test)=mnist_data.load_data()
```
You will notice that there are 60000 samples consisting of images of size 28 × 28. We converted the dataset into 50,000 arrays of size 784. Also, we normalize the dataset by dividing each pixel by 255.
```
X_train= np.reshape(X_train, (X_train.shape[0], X_train.shape[1]*X_train.shape[2]))
X_test= np.reshape(X_test, (X_test.shape[0], X_test.shape[1]*X_test.shape[2]))
X_train=X_train/255
X_test=X_test/255
X_train = X_train.astype('float32')
X_test = X_test.astype('float32')
```
Now we create a model having an input layer of size 784, a hidden layer

CHAPTER 11 AUTOENCODERS

having 512 units, and an output layer of 784 units. We use mean squared loss and Adam optimizer to train the model.
model = tf.keras.Sequential()
model.add(tf.keras.layers.Dense(units=512,activation='sigmoid',input_shape=(784,)))
model.add(tf.keras.layers.Dense(units=784, activation='sigmoid'))
model.compile(loss='MeanSquaredError',optimizer='adam',metrics=['MeanSquaredError'])

It can be observed that there are 804112 trainable parameters.
Model: "sequential_4"

Layer (type)	Output Shape	Param #
dense_6 (Dense)	(None, 512)	401920
dense_7 (Dense)	(None, 784)	402192

Total params: 804112 (3.07 MB)
Trainable params: 804112 (3.07 MB)
Non-trainable params: 0 (0.00 Byte)

Now we train the network through 100 epochs with a batch size of 128.
epochs = 100
history = model.fit(X_train, X_train,epochs=epochs,validation_data=(X_test, X_test),batch_size=128,verbose=2)

Output:

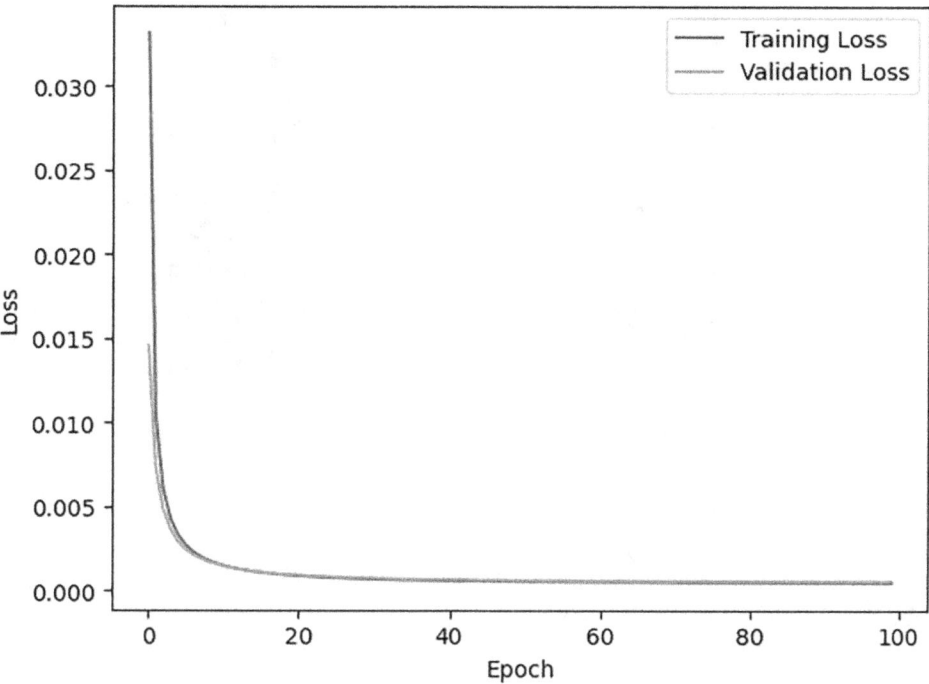

Figure 11-5. *Training and validation loss curves with number of epochs*

The training and validation loss curves are shown in Figure 11-5. Also, some of the reconstructed images are shown in Figure 11-6. The corresponding test images are shown in Figure 11-7.

CHAPTER 11 AUTOENCODERS

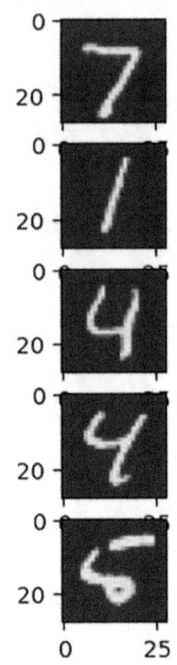

Figure 11-6. *Reconstructed images*

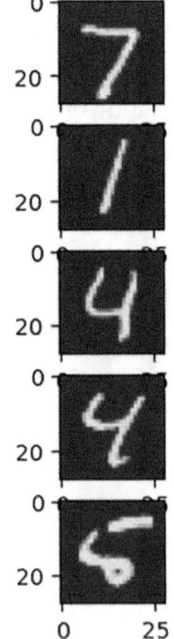
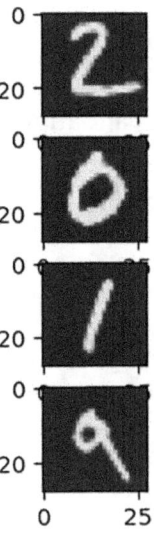

Figure 11-7. *Original test images*

CHAPTER 11　AUTOENCODERS

As far as the MNIST dataset is concerned, very good reconstruction can be obtained using a latent representation of size 512.

Experiment 2

The MNIST dataset is slightly easy to reconstruct as it contains only ten digits. We repeated the experiment with the CIFAR-10 dataset, which also has ten classes, namely:

- Airplane
- Automobile
- Bird
- Cat
- Deer
- Dog
- Frog
- Horse
- Ship
- Truck

Some of the images of this dataset are shown in Figure 11-8.

Figure 11-8. *Images of the CIFAR-10 dataset*

CHAPTER 11 AUTOENCODERS

It can be observed that the images are complex. It is slightly difficult to reconstruct the images using a latent representation. The following experiment reconstructs the images using a latent representation of 512 and a single hidden layer. Some changes have been made in order to make the network learn the hidden representation (Listing 11-2).

Listing 11-2. Reconstructing the CIFAR-10 dataset using an autoencoder

Code:

First of all, the images (50,000 train and 10,000 test) have been converted into grayscale using the following function.

```
def oneDtotwoD(X):
  X1 = []
  for i in range(X.shape[0]):
    img1 = X[i,:,:,:]
    img_gray = 0.2989 * img1[:,:,0] + 0.5870 * img1[:,:,1] + 0.1140 * img1[:,:,2]
    X1.append(img_gray)
    print(len(X1))
  return X1
```

All the images have been flattened and normalized using the following code.

```
X_train= np.reshape(X_train, (X_train.shape[0], X_train.shape[1]*X_train.shape[2]))
X_test= np.reshape(X_test, (X_test.shape[0], X_test.shape[1]*X_test.shape[2]))
X_train=X_train/255
X_test=X_test/255
X_train = X_train.astype('float32')
X_test = X_test.astype('float32')
```

This is followed by the creation of the model having 512 units in the hidden layer and 1024 units in the input and output layer.

```
model = tf.keras.Sequential()
model.add(tf.keras.layers.Dense(units=512,activation='sigmoid',input_shape=(1024,)))
model.add(tf.keras.layers.Dense(units=1024, activation='sigmoid'))
model.compile(loss='MeanSquaredError',optimizer='adam',metrics=['MeanSquaredError'])
```

Output:

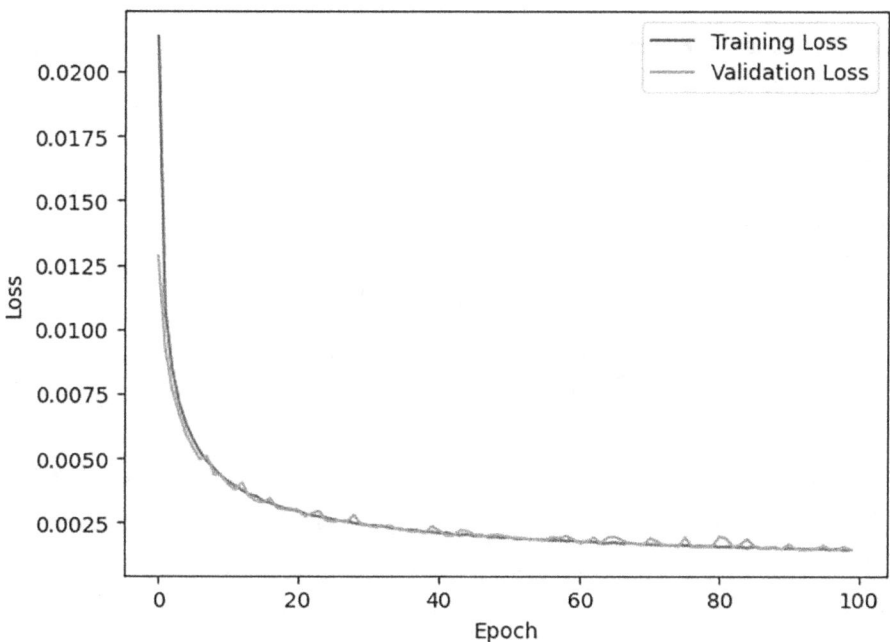

Figure 11-9. *Training and validation loss curves with number of epochs*

The training and validation loss curves are shown in Figure 11-9. The reconstruction here is not as good as in the previous case, which can be inferred by observing the values of the losses (6.02×10^{-4} in the case of the MNIST dataset and 0.001 in the case of the CIFAR-10 dataset).

Finding Latent Representation Using Multiple Layers

It was stated in the previous section that the learning of weights in the case of autoencoders having multiple layers follows a slightly different procedure as compared with a normal dense network. The following code (Listing 11-3) implements the finding of the latent representation of images using multiple layers.

CHAPTER 11   AUTOENCODERS

Listing 11-3. Latent representation using multiple layers

Code:
First of all, we import the necessary libraries to create the model
```
import cv2
import numpy as np
from tensorflow import keras
from tensorflow.keras import layers
```
We then create a function to build autoencoder
```
def build_autoencoder(input_dim, hidden_dim):
  input_layer = layers.Input(shape=(input_dim,))
  encoded = layers.Dense(hidden_dim,activation='relu',name='encoder')
  (input_layer)
  decoded = layers.Dense(input_dim, activation='sigmoid')(encoded)
  autoencoder = keras.Model(input_layer, decoded)
  autoencoder.compile(optimizer='adam', loss='mse')
  autoencoder.summary()
  return autoencoder
```
This is followed by loading the imaging dataset
```
data = np.load('/content/drive/MyDrive/Emotion Detection/X_test_Happy.npy')
print(data.shape)
```
All the imagesare then flattened
```
data = data.reshape((len(data), np.prod(data.shape[1:])))
data.shape
```
We then choose the size of hidden dimensions and train the model
```
hidden_dims = [1024, 512]
encoder_model = None
for i, hidden_dim in enumerate(hidden_dims):
  if i == 0:
    autoencoder = build_autoencoder(data.shape[1], hidden_dim)
  else:
    autoencoder = build_autoencoder(encoder_model.output_shape[1],
    hidden_dim)
  autoencoder.fit(data, data, epochs=10, batch_size=32)
  encoder_model = keras.Model(autoencoder.input,
  autoencoder.get_layer('encoder').output)
```

```
    data = encoder_model.predict(data)
final_encoder = encoder_model
print(data.shape)
```
Output:
Shape of original data having 296 grayscale images of size 224 × 224
(296, 224, 224)
Shape of data after flattening the 296 grayscale images of size 224 × 224
(296, 50176)
Summary of Model 1

Layer (type)	Output Shape	Param #
input_3 (InputLayer)	[(None, 50176)]	0
encoder (Dense)	(None, 1024)	51381248
dense_2 (Dense)	(None, 50176)	51430400

Total params: 102811648 (392.20 MB)
Trainable params: 102811648 (392.20 MB)
Non-trainable params: 0 (0.00 Byte)

Summary of Model 2

Layer (type)	Output Shape	Param #
input_4 (InputLayer)	[(None, 1024)]	0
encoder (Dense)	(None, 512)	524800
dense_3 (Dense)	(None, 1024)	525312

Total params: 1050112 (4.01 MB)
Trainable params: 1050112 (4.01 MB)
Non-trainable params: 0 (0.00 Byte)

Shape of the encoded representation
(296, 512)

Now that we have seen how to find the embedding of hidden layers in a stacked autoencoder, let us move to other variants of autoencoders.

CHAPTER 11   AUTOENCODERS

Variants of Autoencoders

This section discusses some other variants of autoencoders such as sparse, denoising, and variational autoencoders.

Sparse Autoencoder

A sparse autoencoder is a special type of autoencoder that incorporates sparsity during the training process. In section "Types of Autoencoders," we introduced the overcomplete autoencoder in which the hidden layer contains more units than the input layer. This results in the activation of only a small number of neurons. We can achieve this sparsity by adding an extra term to the hidden layer that penalizes the activations. The task can be accomplished using the KL divergence.

$$\text{Loss } J(W) = L(X, \hat{X}) + \lambda \sum_j KL(\rho \| \hat{\rho}_j)$$

where $\sum_j KL(\rho \| \hat{\rho}_j) = \sum_j \left[\rho \log \frac{\rho}{\hat{\rho}_j} + 1 - \rho \log \frac{1-\rho}{1-\hat{\rho}_j} \right]$

and $\hat{\rho}_j = \frac{1}{m} \sum_{i=1}^{m} a_j^h(x_i)$

Here, a_j^h = activation of the jth neuron in hidden layer h.

Subject to constraint: $\hat{\rho}_j = \rho$

The characteristics of sparse autoencoders are as follows:

- Sparse autoencoders implement sparsity during the training process.

- They can learn features even when the hidden layer has more neurons than the input layer.

- A sparsity constraint in the hidden layer ensures that only a small portion of neurons are activated.

- An additional term is included in the loss function to penalize hidden layer activations, pushing them toward zero. As stated earlier, this can be implemented using L1 regularization or KL divergence.

- By enforcing sparsity, the network learns to capture the most significant features.

Denoising Autoencoder

Denoising autoencoders are a special type of autoencoder that can remove noise from data. They have an architecture similar to regular autoencoders, consisting of an encoder and a decoder. The encoder processes the noised version of input data and converts it into a lower-dimensional representation. This compressed representation captures the essential, noise-free features of the data. The decoder then receives this encoded representation and tries to reconstruct an uncorrupted version of the original input. This process augments the ability of a network to capture the underlying patterns of the data, thus making it more robust to noise and improving the model's performance. Denoising autoencoders are used in various applications such as image, signal, and text denoising.

Variational Autoencoder

A variational autoencoder (VAE) is an autoencoder as it is designed to compress high-dimensional input data into a smaller representation. Whereas a typical autoencoder maps the input data to a latent vector, a VAE on the other hand maps the input to the parameters of a probability distribution, namely, (a) the mean and (b) the variance. It is particularly effective for image generation.

Conclusion

This chapter presents a brief introduction to autoencoders. Autoencoding entails training a network to replicate its input as its output, thus learning the latent representation of the input. This process is important for developing embeddings that help in information retrieval. Autoencoders can be viewed as a type of lossy compression, wherein the network identifies the essential attributes of the input. Depending on the size of the hidden layer, autoencoders can be under-complete, with a hidden layer size smaller than the input layer, or over-complete, with larger hidden layers. A stacked autoencoder includes multiple hidden layers. Furthermore, these networks can be trained to denoise input by using corrupted instances as input. The chapter also gives a very brief introduction to a variational autoencoder that functions as a generative model that can generate samples from the learned latent space.

CHAPTER 11 AUTOENCODERS

Exercises
Multiple-Choice Questions

1. What is the primary purpose of an autoencoder?

    a. To classify input data using maximum margin classifier

    b. To replicate its input to its output

    c. To predict future data points using sequence modelling

    d. To cluster similar data points

2. How does an autoencoder support information retrieval?

    a. By generating new data points

    b. By learning embeddings

    c. By clustering data

    d. By reducing noise in data

3. What is an autoencoder trained to learn?

    a. The function that maps input to itself

    b. The difference between input and output

    c. The classification boundaries

    d. The regression function

4. An autoencoder can be thought of as which of the following?

    a. Lossless compression of input

    b. Lossy compression of input

    c. Generative modeling

    d. Predictive modeling

5. What must an autoencoder identify to reproduce inputs closely?

    a. The noise in the data

    b. The important attributes of inputs

c. The future values of the data

d. The clustering structure

6. What is a characteristic of under-complete autoencoders?

   a. Hidden layer size larger than input layer size

   b. Hidden layer size equal to input layer size

   c. Hidden layer size smaller than input layer size

   d. No hidden layers

7. What defines an over-complete autoencoder?

   a. Much larger hidden layer sizes

   b. Hidden layer size equal to input layer size

   c. Hidden layer size smaller than input layer size

   d. No hidden layers

8. What is a stacked autoencoder?

   a. An autoencoder with a single hidden layer

   b. An autoencoder with multiple hidden layers

   c. An autoencoder without hidden layers

   d. An autoencoder with a large input layer

9. How can an autoencoder be trained to learn to denoise input?

   a. By using only clean data as input

   b. By giving input and a corrupted instance and targeting the uncorrupted instance

   c. By using a larger hidden layer size

   d. By clustering the input data

10. What is a variational autoencoder (VAE)?

    a. An autoencoder without a hidden layer

    b. An autoencoder that is also a generative model

    c. An autoencoder with a larger input layer size

    d. An autoencoder that uses supervised learning

Theory

1. What is an autoencoder? How does a multi-layer autoencoder learn?

2. What is a denoising autoencoder? Write an algorithm to remove noise from a set of images from a particular distribution using this network.

3. What is a variational autoencoder? Write the objective function of VAE and explain its importance.

4. Compare an autoencoder with Principal Component Analysis.

Applications

Hari wants to develop a monument identification application that can identify all the major monuments in Delhi. The idea is that if a tourist clicks the picture of a monument using the app, the app should be able to classify the monument and show its details. To develop such an app, he gathered 3000 images of each monument from the Internet.

He tried using conventional feature extraction methods but was not very successful. Can you help him accomplish the task using autoencoders?

Can autoencoders help him denoise some images of the same monuments clicked by the phones of his employees? Explain how this can be done.

Finally, he wants to generate photos of new kinds of monuments using the app. Can you help him accomplish this task using VAE?

CHAPTER 12

Introduction to Generative Models

Introduction

Since we have reached the end of our journey, let us contemplate what we were expecting when we started. The goal was to be able to develop models that can do image- and sequence-related tasks efficiently and effectively. We now know that DL models can help us classify images and text. Chapter 6 to Chapter 10 of this book focus on the convolutional and sequential models that help us accomplish such tasks. We also learned to develop models that can accomplish slightly complex tasks like next character generation and encoding of an image. Let us now focus on more complex tasks and explore the fundamentals of generative models. Generative models not only help us carry out supervised and unsupervised learning tasks, studied so far, but also help us generate new data from a particular distribution. One of the glaring examples of generative models is ChatGPT, which has disrupted the field. It is based on transformers. This chapter introduces transformers. But before diving into transformers, let us have a basic idea of Hopfield Networks and Boltzmann Machines.

Hopfield Networks

If you hear the song "Turn! Turn! Turn!", what comes to your mind? Perhaps The Byrds, the band, or Forrest Gump or the first eight verses of the third chapter of the biblical Book of Ecclesiastes? In your lifetime you must have heard a lot of songs. In spite of that, on hearing a few lines of a famous song, the whole song, the image associated with it, and the source come to your mind. Being a computer science student, what do you think

CHAPTER 12 INTRODUCTION TO GENERATIVE MODELS

goes in our brain that helps us associate the few lines with the complete description or perhaps the partial one? Is it the database of all the songs that is created in our mind followed by some type of search that associates a pattern with a particular song? Perhaps the answer is a no!

These search strategies cannot work and produce answers in microseconds, so what exactly happens? The answer lies in the ability of a particular object to attain the state of minimum energy, that is, pure physics. The computational model that implements this strategy was given by John Hopfield in 1982. His idea was based on the strategy followed by proteins to attain a stable structure, one that minimizes their energy. This model is referred to as the **Hopfield Network**.

Assume that we store a pattern consisting of $\{x_1, x_2, x_3 .. x_n\}$. Also, assume that each of these x_is can either be +1 or -1. The interaction between them can be depicted by a graph having x_is as the vertices and w_{ij} as the weight between patterns x_i and x_j. To keep things simple, let us assume that the graph so formed is a unidirectional graph. For example, consider the graph shown in Figure 12-1 having three vertices x_1, x_2, and x_3 and weights w_{12}, w_{32}, and w_{31}.

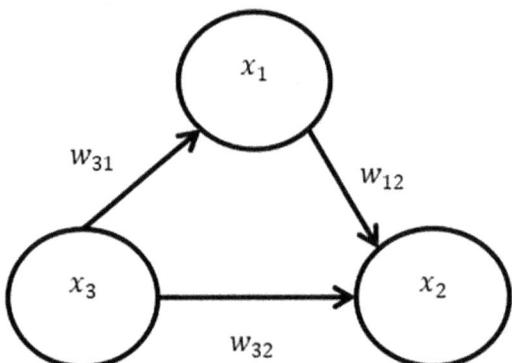

Figure 12-1. *A Hopfield Network consisting of patterns x_1, x_2, x_3*

Note that, in this network, if w_{ij} is greater than 0, then the connection between them is considered as exhibitory; likewise, if the weights between them are less than 0, then it is considered as inhibitory.

To begin with, let us consider only two patterns x_1 and x_2, both of which either can be +1 or -1. Then Table 12-1 shows the sign of weights between them.

CHAPTER 12 INTRODUCTION TO GENERATIVE MODELS

Table 12-1. *Finding Weights When the Values of x_1s Are Given*

X_1	X_2	W_{12}
+1	+1	>0
+1	-1	<0
-1	+1	<0
-1	-1	>0

This means that if both x_i and x_j have the same sign, then the weight is positive; else, the weight is negative. Does this remind of you anything? This is Hebb's rule:

"Neurons that wire together fire together"

This leads us to a factor that is to be maximized if the whole configuration is to become stable, which is $\sum w_{ij} x_i x_j$. This means that the following quantity needs to be minimized:

$$Energy = -\sum w_{ij} x_i x_j$$

This may be referred to as energy. The Hopfield Network aims to minimize this energy. In order to achieve a stable configuration on giving a particular pattern, we need to find out the values of x_i and x_j to make the configuration stable and the corresponding weights. For finding out the values of x_is and the weights, the following strategy may be applied (Figure 12-2).

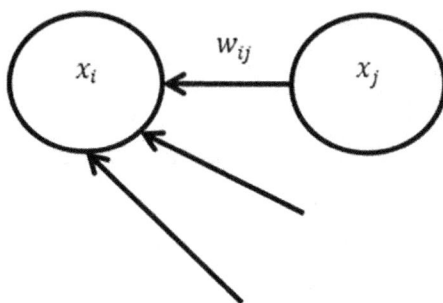

Figure 12-2. *Finding the updated values of x_i*

Task 1: Finding the values of x_is

- Find $S_i = \sum x_j w_{ij}$
- If $S_i > 0$ then set $x_i = 1$; else, set $x_i = -1$.
- Repeat the process for all x_is, and continue repeating till a stable state is reached.

Task 2: Finding weights

- For a given pattern y_i, set $w_{ij} = y_i y_j$.
- As we need to minimize $-\sum w_{ij} y_i y_j$ and the minimum value is attained at $w_{ij} = y_i y_j$

This way the new weights can be found. The readers interested in derivation may refer to the references given at the end of the chapter. Let us now have a look at a machine that can model binary data.

Boltzmann Machines

Assume that you are working in a control room of a factory and all the buttons there can only be in one of the two states: on or off (0 or 1). The control room's configuration can be defined in terms of the state of each of these buttons. It is important to find out if the configuration is problematic, as something can seriously go wrong in such cases. Let us formally state the problem:

Given a set of binary variables $\{x_1, x_2, ...x_m\}$, we need to find out if a vector of length m depicting the state of each of these variables presents a condition of anomaly.

So we need to develop a machine that is able to model the binary data. One of the ways of doing so is to use a Boltzmann Machine (BM). A Boltzmann Machine can model binary data [2]. Using this machine, we can find if a given vector belongs to a particular distribution. Likewise, if you develop a few such machines, you can, with the help of Bayes' theorem, find if the vector came from a particular distribution. These machines, when modeled on a normal state, can also help us find out about unusual behavior.

Let's consider a scenario wherein we need to generate data from a binary distribution. To be able to do so, we need to find the latent variables, followed by developing a network with hidden states and visible states. We first use the prior distribution and choose the hidden states and then find the visible state from the

CHAPTER 12 INTRODUCTION TO GENERATIVE MODELS

conditional distribution. However, Boltzmann Machines do not work in this way. In these machines, the energy of the joint configuration is proportional to the probability P(v, h), where v is the visible state and h is the hidden state.

The probability of a visible state here is

$$P(v) = \sum_h P(h) \times P(v/h)$$

As per Reference [1], the energy $E(v, h)$ is given by the formula

$$E(v,h) = -\left(\sum_i v_i b_i + \sum_j h_j b_j + \sum_{i,j} h_i h_j w_{ij} + \sum_{i,j} v_i v_j w_{ij} + \sum_{i,j} v_i h_j w_{ij}\right)$$

The value of $P(v)$ can be calculated using the following formula:

$$P(v,h) = e^{-E(v,h)} / \sum_{x,y} e^{-E(x,y)}$$

And finally, $P(v)$ can be calculated using the following formula:

$$P(v) = \sum_h e^{-E(v,h)} / \sum_{x,y} e^{-E(x,y)}$$

To understand how probability distribution of various visible states is derived in a Boltzmann Machine, consider the following example that follows.

In Figure 12-3 we have three hidden states and three visible states. The weight between h_1 and h_2 is 2; that between h_1 and v_1 is 3; that between h_2 and v_2 is -1; that between h_2 and h_3 is 1; and that between h_3 and v_3 is 2. To find the probabilities of various states, the following steps must be followed.

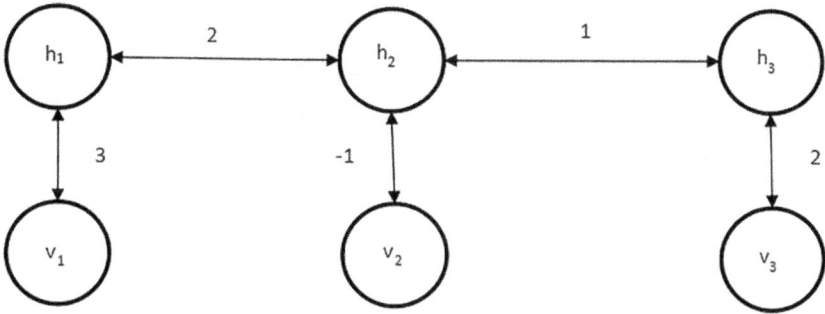

Figure 12-3. *An example of a Boltzmann Machine*

CHAPTER 12　INTRODUCTION TO GENERATIVE MODELS

Step 1: We enlist all possible permutations of binary variable (h_1 h_2 h_3), which are eight values. Note that (v_1 v_2 v_3) can also have eight values, and thus we have 64 combinations in total (Table 12-2).

Table 12-2. Sum of the Possible Combinations for Visible and Hidden States

v_1 v_2 v_3	h_1 h_2 h_3
000	000
001	000
010	000
011	000
100	000
101	000
110	000
111	000
000	001
...	...
111	111

Step 2: This is followed by calculating E for each of the 64 combinations obtained. For instance, take the case when (v_1 v_2 v_3) are (1 1 0) and (h_1 h_2 h_3) are (0 1 0), respectively.

Assume that the values of v_1, v_2, and v_3 are 1,1, and 0 and those of h_1, h_2, and h_3 are 0, 1, and 0.

For the sake of simplicity, assume all the biases are 0 so $\sum w_i b_i$ and $\sum h_k b_k$ become 0. Hence, we are left with $\sum v_i h_k w_{ik}$ and $\sum h_k h_l w_{kl}$. Note that, visible states, that is, (v_1 v_2 v_3), are not connected with each other (Figure 12-4).

CHAPTER 12 INTRODUCTION TO GENERATIVE MODELS

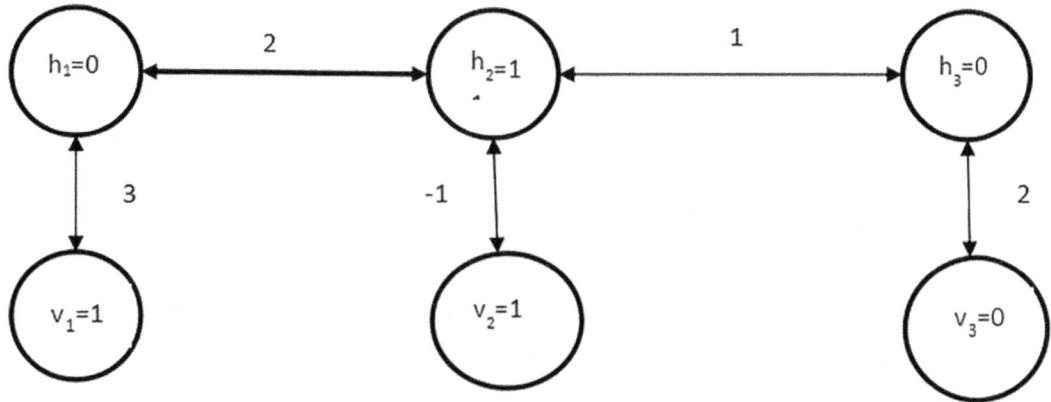

Figure 12-4. *Boltzmann Machine along with the state inputs*

To calculate $\sum v_i h_k w_{ik}$ and $\sum h_k h_l w_{kl}$, we get

$$= v_1 h_1 w_{11} + v_2 h_2 w_{22} + v_3 h_3 w_{33} + h_1 h_2 w_{12} + h_2 h_3 w_{23}$$

$$= 0 + (-1) + 0 + 0 + 0$$

$$= -1$$

Since $-E = -1$ so $e^{-E} = e^{-1}$.

As another example, consider another case when $(v_1\ v_2\ v_3)$ are $(1\ 1\ 1)$ and $(h_1\ h_2\ h_3)$ are $(1\ 1\ 1)$, respectively. Assume that the values of v_1, v_2, and v_3 are 1, 1, and 1 and those of h_1, h_2, and h_3 are 1, 1, and 1.

On calculating $\sum v_i h_k w_{ik}$ and $\sum h_k h_l w_{kl}$, we get

$$= v_1 h_1 w_{11} + v_2 h_2 w_{22} + v_3 h_3 w_{33} + h_1 h_2 w_{12} + h_2 h_3 w_{23}$$

$$= 3 + (-1) + 2 + 2 + 1$$

$$= 7$$

We now know $-E = -7$, so $e^{-E} = e^7$.

This way we can calculate the value of all e^{-E} for all the combinations mentioned above and find the sum. Now, we divide each e^{-E} with the sum calculated above to get the probability of each combination.

Now, consider a situation wherein you have a lot of visible and hidden states. In such cases enumerating all the possible combinations and then finding out the probability of all the visible states becomes computationally difficult. To handle this problem, Boltzmann Machines were proposed.

CHAPTER 12   INTRODUCTION TO GENERATIVE MODELS

Boltzmann Machines (BMs) and Restricted Boltzmann Machines (RBMs) are both types of stochastic Neural Networks, but they have significant differences in their structures and applications.

The Boltzmann Machines are fully connected networks. The Restricted Boltzmann Machine has a bipartite graph structure. The former are computationally demanding, whereas the learning in the latter is simple and is done through Contrastive Divergence (CD). The Boltzmann Machines are generally used for solving optimization problems, whereas the latter is used for feature learning and dimensionality reduction. The latter is practical and it is easy to train with larger datasets. Having seen the basis of Hopfield and Boltzmann machines, let us now move to transformers.

A Gentle Introduction to Transformers

This section is based on an original research paper called "Attention is all you need" by Vaswani et al. and its explanation on the New York University website by Chinmay Hegde.

The Large Language Models (LLMs) have become extremely popular for the past few years, particularly with the advent of ChatGPT. These models can perform various tasks like

   i. Summarization of text

   ii. Generating new text

   iii. Correcting the existing ones

   iv. Translation (to some extent)

In this book, we have already studied Recurrent Neural Networks (RNNs), which can deal with sequences. We have already seen applications like Sentiment Analysis, Named Entity Recognition, generating the next character, etc. using RNNs and their variants. However, RNNs do not perform well on tasks like language translation.

Assume that you aim to develop an application that converts English to Marathi. Your application takes an input sentence in English and generates a sentence in Marathi. For example, if the input sentence is

"I eat rice"

then the output sentence should be

"मी भातखातो"

Note that the second word in the source language is "eat," whereas it is equivalent to the third position in the target sentence. This is called misalignment, and RNNs do not handle this problem gracefully. Likewise, consider another sentence:

"I am a good boy"

then the corresponding sentence in Marathi will be

"मी चांगला मुलगा आहे"

Note that the sentence in the source language contains five words, whereas the target sentence has four words. In such cases, the number of words in the source language may not be the same as in the target language. In machine translation

  i. The number of words in the source language may not be the same as in the target language for a particular sentence.

  ii. There can be misalignment.

To solve this problem, the following approaches can be employed. Instead of creating a word-level RNN, we can make a sentence-level RNN. However, this approach would not work well because, for a given combination of words, there can be many sentences, and the model might not understand the context and placement accurately.

The second option is to create an encoder–decoder-type architecture, as explained in Chapter 9 on RNN. Here, we will focus on another solution to this problem, which forms the basis of modern-day ChatGPT.

An Introduction to Self-Attention

Assume that we have a sentence(X) consisting of some words (x_i) each having dimension (d). The output will be a sentence Y consisting of y_i, also a d-dimensional vector. The sentence contains the set $y_1, y_2, y_3, ..., y_n$ such that

$$y_i = \sum_{j=1}^{n} x_j w_{ij}$$

where w_{ij} is the weight corresponding to i^{th} vector in the output and the jth vector in the input. Also, w_{ij} is row normalized. Here, the initial weights are chosen as

$$w_{ij} = x_i^T \cdot x_j$$

and then we apply the softmax function to find W_{ij}:

$$W_{ij} = \frac{e^{w_{ij}}}{\sum_{k=1}^{n} e^{w_{ik}}}$$

CHAPTER 12   INTRODUCTION TO GENERATIVE MODELS

In such models, a single input is generally mapped to a set of outputs. This model is capable of considering all the units of the input. Note that the embeddings of each word can be learned by some conventional method or a Neural Network. This model is capable of handling many of the issues stated above; however, some issues are still to be addressed (Figure 12-5):

i. In such type of model, the system input, say x_i, is multiplied with all the other vectors to build the sequence of weights:

$$w_{i1} = x_i \cdot x_1 \; w_{i2} = x_i \cdot x_2 \ldots w_{ij} = x_i \cdot x_j$$

This role is called "Query."

ii. It is then compared with every other point to get the weight of the output

$$w_{ji} = x_1^T \cdot x_i$$

$$w_{ji} = x_2^T \cdot x_i \ldots w_{ji} = x_j^T \cdot x_i$$

used for finding y_i. This role is called "Key."

iii. Then the outputs $y_1, y_2, y_3, \ldots, y_n$ are synthesized. This role is called "Value."

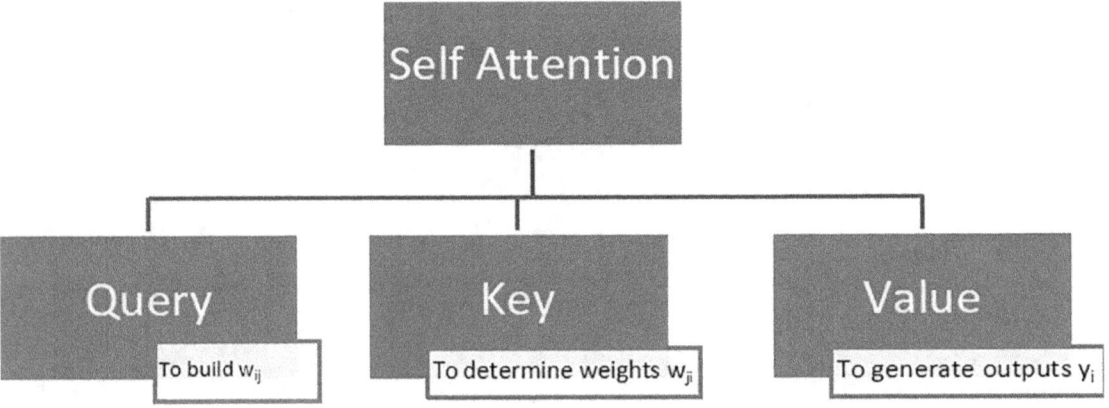

Figure 12-5. *Components of self-attention*

CHAPTER 12 INTRODUCTION TO GENERATIVE MODELS

$$Q_i = W_q \times X_i \quad K_i = W_k \times X_i \quad V_i = W_v \times X_i \quad w_{ij} = \frac{Q_i^T \times K_j}{\sqrt{d}} \quad W_{ij} = Softmax(w_{ij})$$

$$= \frac{e^{w_{ij}}}{\sum_{k=1}^{n} e^{w_{ik}}} \quad y_i = \sum_{j=1}^{n} W_{ij} \times V_j$$

In addition to the above, we can also use multi-head self-attention. Interested readers may refer to the references given at the end of this chapter for understanding multi-head self-attention.

The Transformer

A transformer consists of a self-attention block followed by a layer of normalization, then a Multi-layer Perceptron (MLP), and then another self-attention block as shown in Figure 12-6.

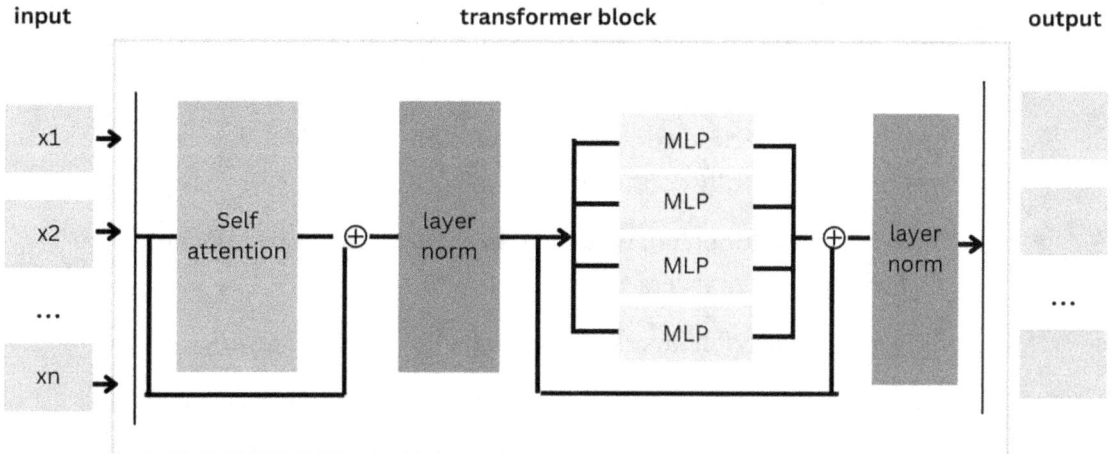

Figure 12-6. *Transformer architecture*

Transformers have many advantages:

  i. We can easily club multiple transformers together.
  ii. They use a fully feed-forward architecture for parallelization.
  iii. They support standard backpropagation for training.
  iv. Transformers are highly scalable.
  v. They handle variable-length sequences efficiently.

CHAPTER 12   INTRODUCTION TO GENERATIVE MODELS

Conclusion

This chapter covers three important models: Hopfield Networks, Boltzmann Machines, and the self-attention mechanism. These models are the basis of generative models and modern pattern recognition techniques. Each topic is introduced with examples. For the readers looking for a more in-depth information, references are provided at the end of the chapter. In addition to the above, it is worth noting that self-attention mechanisms and transformers are the technologies behind ChatGPT.

Exercise

Multiple-Choice Questions

1. What is the most important structural difference between Boltzmann Machines (BMs) and Restricted Boltzmann Machines (RBMs)?

    a. BMs have connections only between visible and hidden layers, while RBMs are fully connected.

    b. BMs are fully connected, while RBMs have connections only between visible and hidden layers.

    c. BMs have no hidden layers, while RBMs have hidden layers.

    d. BMs use supervised learning, while RBMs use unsupervised learning.

2. Which learning algorithm is commonly used to train Restricted Boltzmann Machines (RBMs)?

    a. Stochastic gradient descent

    b. Backpropagation

    c. Contrastive Divergence

    d. Gradient Boosting

CHAPTER 12   INTRODUCTION TO GENERATIVE MODELS

3. In terms of complexity and training, how do Boltzmann Machines (BMs) compare with Restricted Boltzmann Machines (RBMs)?

    a. BMs are simpler and faster to train compared with RBMs.

    b. BMs and RBMs have the same complexity and training speed.

    c. BMs are more complex and harder to train compared with RBMs.

    d. RBMs are more complex and harder to train compared with BMs.

4. Which of the following is a common application of Restricted Boltzmann Machines (RBMs)?

    a. Solving optimization problems

    b. Feature learning and dimensionality reduction

    c. Image classification

    d. Natural language processing

5. What type of network is a Hopfield Network?

    a. Feed-Forward Neural Network

    b. Recurrent Neural Network

    c. Convolutional Neural Network

    d. Generative Adversarial Network

6. In a Hopfield Network, what kind of values do the neurons typically hold?

    a. Continuous values between 0 and 1

    b. Continuous values between -1 and 1

    c. Binary values (0 or 1)

    d. Binary values (-1 or 1)

7. What is a primary application of Hopfield Networks?

    a. Supervised learning

    b. Image classification

CHAPTER 12 INTRODUCTION TO GENERATIVE MODELS

  c. Pattern recognition and associative memory

  d. Natural language processing

8. Which of the following happens with the energy function of a Hopfield Network?

  a. It increases as the network stabilizes.

  b. It decreases as the network stabilizes.

  c. It remains constant as the network stabilizes.

  d. It is not defined for Hopfield Networks.

9. What is the most important purpose of the self-attention mechanism in Neural Networks?

  a. To reduce the dimensionality of the input data

  b. To allow the network to focus on different parts of the input sequence when processing each element

  c. To improve the computational efficiency of the network

  d. To enable the network to perform unsupervised learning

10. In the self-attention mechanism, what are the three main components that are derived from the input vectors?

  a. Inputs, hidden states, and outputs

  b. Weights, biases, and activations

  c. Queries, keys, and values

  d. Layers, nodes, and edges

Theory

1. Explain the terms key, value, and query vis-à-vis the self-attention mechanism.

2. Explain how a Boltzmann Machine can be used to complete a given partial image.

3. Explain the idea of a Hopfield Network. Explain any three applications of such networks.

4. Explain the structure of transformers.

References

[1] *Boltzmann Machines.* https://classes.engr.oregonstate.edu/eecs/winter2020/cs536/slides/boltzmanmachines.4pp.pdf

[2] Hinton, G. *CSC321: Introduction to Neural Networks and Machine Learning Lecture 18 Learning Boltzmann Machines.* https://www.cs.toronto.edu/~hinton/csc321/notes/lec18.pdf

[3] Hinton, G., Srivastava, N., Swersky, K., Tieleman, T. & Mohamed, A. Neural networks for machine learning. In *Lecture 11a* (n.d.). https://www.cs.toronto.edu/~hinton/coursera/lecture11/lec11.pdf

[4] *MIT OpenCourseWare.* MIT OpenCourseWare. https://ocw.mit.edu/courses/9-40-introduction-to-neural-computation-spring-2018/resources/mit9_40s18_lec20/

[5] *Lecture 23: Associative memory & Hopfield Networks.* https://gyansanchay.csjmu.ac.in/wp-content/uploads/2022/02/AssociativeMemoryHopfieldNetworks.pdf

[6] Rossa, C. Actuators and power electronics. In *METE 3100U.* https://www.biomechatronics.ca/teaching/ape/notes/Lecture_3.pdf

APPENDIX A

Classifying The Simpsons Characters

This appendix aims to develop a Convolutional Neural Network (CNN) model for the classification of characters of *The Simpsons*. In total 3000 images of 10 characters have been extracted from the original source (https://www.kaggle.com/datasets/alexattia/the-simpsons-characters-dataset). Each class has 300 images. Figure A-1 shows an instance of each class. The images have been resized to (224, 224, 3) and normalized using min–max normalization.

Figure A-1. *An instance of each of the ten classes*

```
X_train, X_test = X_train / 255.0, X_test / 255.0
print(X_train.shape, X_test.shape)
```

APPENDIX A CLASSIFYING THE SIMPSONS CHARACTERS

A CNN called *Model_1* has been developed by creating a sequential model consisting of alternate convolutional and pooling layers (three pairs) followed by two dense layers and the softmax layers:

```
Model_1 = models.Sequential()
Model_1.add(layers.Conv2D(16, (5,5), activation='relu', input_shape=(224, 224, 3)))
Model_1.add(layers.MaxPooling2D((2, 2)))
Model_1.add(layers.Conv2D(32, (3, 3), activation='relu'))
Model_1.add(layers.MaxPooling2D((2, 2)))
Model_1.add(layers.Conv2D(64, (3, 3), activation='relu'))
Model_1.add(layers.MaxPooling2D((2, 2)))
Model_1.add(layers.Flatten())
Model_1.add(layers.Dense(128, activation='relu'))
Model_1.add(layers.Dense(64, activation='relu'))
Model_1.add(layers.Dense(10, activation='softmax'))
Model_1.compile(optimizer='adam',loss='sparse_categorical_crossentropy',metrics=['accuracy'])
Model_1.summary()
```

The summary of the model is as follows:

Model: "Model_1"

Layer (type)	Output Shape	Param #
conv2d (Conv2D)	(None, 220, 220, 16)	1216
max_pooling2d (MaxPooling2D)	(None, 110, 110, 16)	0
conv2d_1 (Conv2D)	(None, 108, 108, 32)	4640
max_pooling2d_1 (MaxPooling2D)	(None, 54, 54, 32)	0
conv2d_2 (Conv2D)	(None, 52, 52, 64)	18496
max_pooling2d_2 (MaxPooling2D)	(None, 26, 26, 64)	0

```
flatten (Flatten)           (None, 43264)              0
dense (Dense)               (None, 128)                5537920
dense_1 (Dense)             (None, 64)                 8256
dense_2 (Dense)             (None, 10)                 650
=================================================================
Total params: 5571178 (21.25 MB)
Trainable params: 5571178 (21.25 MB)
Non-trainable params: 0 (0.00 Byte)
```

The model is compiled with the **Adam** optimizer using sparse **categorical cross-entropy**. It was noted that the training accuracy reached 100% after 25 epochs, while the validation accuracy for the same 25 epochs was 61.33%. The model is recompiled with 100 epochs, and similar results are obtained, which indicates overfitting:

```
batch_size = 64
history_batch = Model_1.fit(X_train, y_train, epochs=25, batch_size=batch_size, validation_data=(X_test, y_test))
plt.plot(history_batch.history['loss'], label='Batch Training Loss')
plt.plot(history_batch.history['val_loss'], label='Batch Validation Loss')
plt.title('Batch Training and Validation Loss')
plt.xlabel('Epochs')
plt.ylabel('Loss')
plt.legend()
plt.show()
```

The variation in the loss with the number of epochs is shown in Figure A-2.

APPENDIX A   CLASSIFYING THE SIMPSONS CHARACTERS

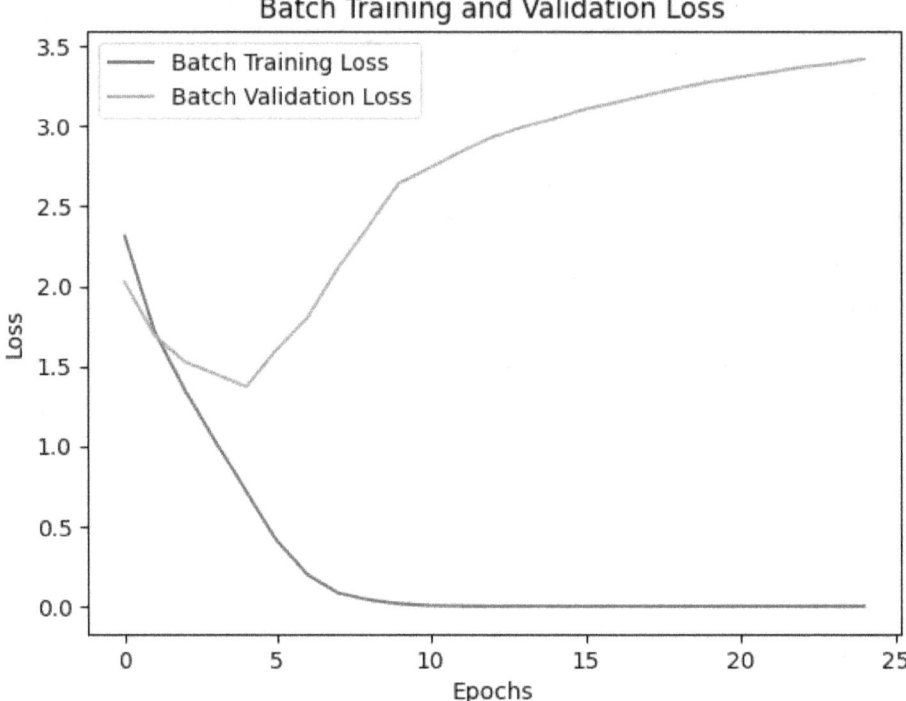

Figure A-2. *Loss curve for Model 1*

Except for the above, two models were created and trained. The summary of the models along with their variation of loss with the number of epochs is shown in Figures A-3 and A-4.

Model: "Model_2"

```
_____
Layer (type)                 Output Shape              Param #
=================================================================
conv2d_3 (Conv2D)            (None, 220, 220, 16)      1216
max_pooling2d_3 (MaxPoolin   (None, 110, 110, 16)      0
g2D)
conv2d_4 (Conv2D)            (None, 108, 108, 32)      4640
max_pooling2d_4 (MaxPoolin   (None, 54, 54, 32)        0
g2D)
conv2d_5 (Conv2D)            (None, 52, 52, 64)        18496
max_pooling2d_5 (MaxPoolin   (None, 26, 26, 64)        0
g2D)
```

flatten_1 (Flatten)	(None, 43264)	0
dropout (Dropout)	(None, 43264)	0
dense_3 (Dense)	(None, 128)	5537920
dense_4 (Dense)	(None, 64)	8256
dense_5 (Dense)	(None, 10)	650

===

Total params: 5571178 (21.25 MB)
Trainable params: 5571178 (21.25 MB)
Non-trainable params: 0 (0.00 Byte)

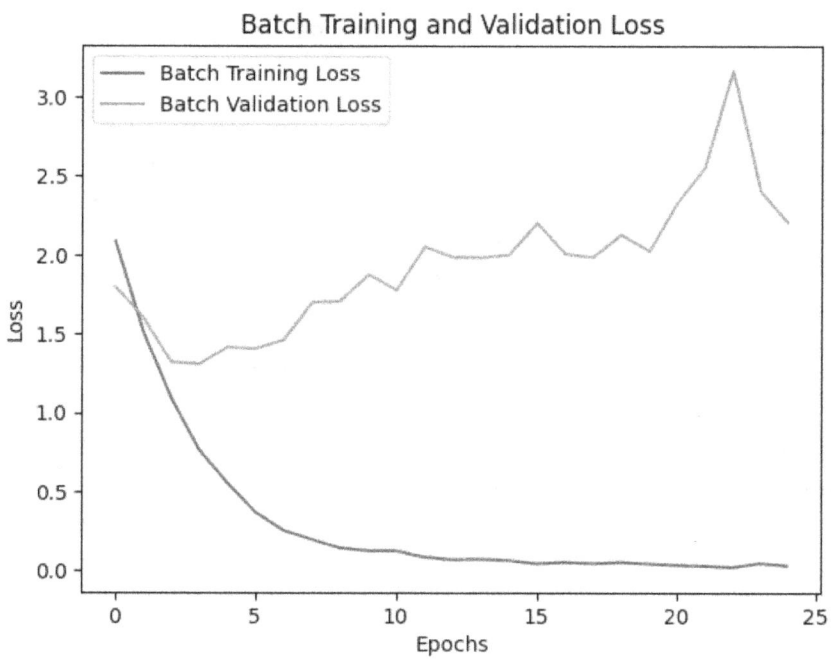

Figure A-3. Loss curve for Model 2

APPENDIX A CLASSIFYING THE SIMPSONS CHARACTERS

Model: "Model_3"

Layer (type)	Output Shape	Param #
conv2d_6 (Conv2D)	(None, 220, 220, 16)	1216
max_pooling2d_6 (MaxPooling2D)	(None, 110, 110, 16)	0
conv2d_7 (Conv2D)	(None, 108, 108, 32)	4640
max_pooling2d_7 (MaxPooling2D)	(None, 54, 54, 32)	0
conv2d_8 (Conv2D)	(None, 52, 52, 64)	18496
max_pooling2d_8 (MaxPooling2D)	(None, 26, 26, 64)	0
flatten_2 (Flatten)	(None, 43264)	0
dropout_1 (Dropout)	(None, 43264)	0
dense_6 (Dense)	(None, 128)	5537920
dropout_2 (Dropout)	(None, 128)	0
dense_7 (Dense)	(None, 64)	8256
dense_8 (Dense)	(None, 10)	650

Total params: 5571178 (21.25 MB)
Trainable params: 5571178 (21.25 MB)
Non-trainable params: 0 (0.00 Byte)

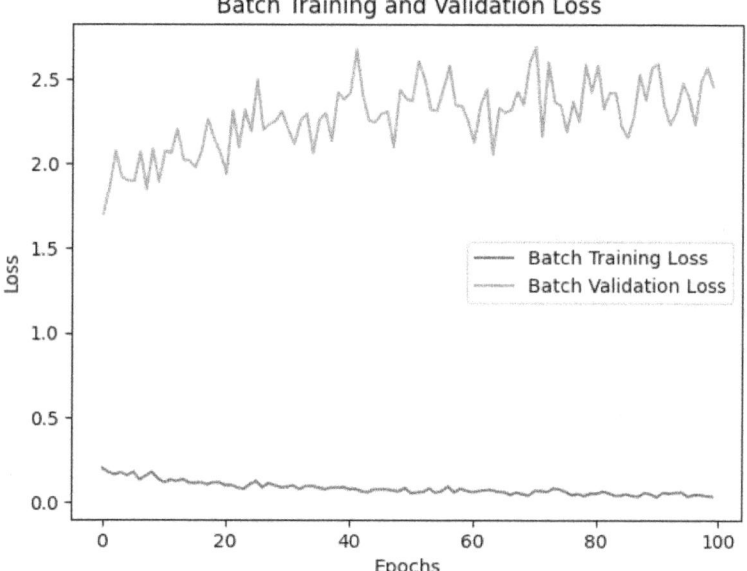

Figure A-4. *Loss curve for Model 3*

The reader is expected to apply the techniques studied in Chapter 5 to handle overfitting. The next appendix draws a bounding box around the faces in an image. The reader is encouraged to find out if the technique works for cartoons also. If not, can you guess the reason?

APPENDIX B

Face Detection

Introduction

This appendix introduces a pretrained model in **Keras** for the detection and classification of faces. The model is Multi-task Cascaded Convolutional Neural Network. The code presented in Listing B-1 requires you to install **MTCNN**, assuming that you have already installed **Matplotlib** and **Keras.** This appendix draws a bounding box around the face in a picture that contains a single face and also a picture that contains multiple faces as shown in Figures B-1 and B-2.

We will read an image using **Matplotlib** and then create an instance of **MTCNN**. We then use the *detect_faces* function for finding out the faces; this is followed by extracting individual faces and extracting patches from the original image using **Matplotlib**.

Listing B-1. Face detection using MTCNN

```
from matplotlib import pyplot as plt
from matplotlib.patches import Rectangle
from mtcnn.mtcnn import MTCNN
img_arr_1 = plt.imread('/content/Image_1.jpg')
img_arr_2 = plt.imread('/content/Image_2.jpg')
detector = MTCNN()
face_images_1 = detector.detect_faces(img_arr_1)
face_images_2 = detector.detect_faces(img_arr_2)
```

APPENDIX B FACE DETECTION

We then create a function called ***find_face*** in which we read an image and for each face in the image we draw a box around it. We can extract the face and carry out further analysis if required:

```
def find_faces(image_path, img_arr):
  image = plt.imread(image_path)
  plt.imshow(image)
  ax = plt.gca()
  for face in img_arr:
    x, y, width, height = face['box']
    print(x, y, width, height)
    face_boundary = Rectangle((x, y), width, height,
                       fill=False, color='red')
    ax.add_patch(face_boundary)
  plt.show()
find_faces('/content/Image_1.jpg',face_images_1)
find_faces('/content/Image_2.jpg',face_images_2)
```

Output:

Figure B-1. *Face detection from an image containing a single face*

Figure B-2. *Face detection from an image containing multiple faces*

The reader may test the above using various pictures obtained from different sources and find out if the model works for animated pictures.

APPENDIX C

Sentiment Classification Revisited

Introduction

This appendix classifies the given sentences according to the sentiments. It utilizes the Twitter US Airline Sentiment dataset from Kaggle. This dataset contains tweets about US airlines and their sentiments (positive, neutral, negative). The dataset is downloaded from Kaggle (https://www.kaggle.com/datasets/crowdflower/twitter-airline-sentiment?resource=download).

The following experiment (Listing C-1) classifies the sentiment of tweets from the Twitter US Airline Sentiment dataset into positive, neutral, or negative category using the variants of Recurrent Neural Network (RNN) architectures. The dataset is first preprocessed by selecting relevant columns, encoding the sentiment labels, tokenizing the text, and padding the sequences. Five different models were created as follows:

Model 1: Simple RNN with a single layer having 64 units

Model 2: Bidirectional RNN with a single layer having 64 units

Model 3: GRU with a single layer having 64 units

Model 4: LSTM with a single layer having 64 units

Model 5: Bidirectional LSTM with a single layer having 64 units

Each model is compiled with the **Adam** optimizer and **sparse categorical cross-entropy** loss and trained for ten epochs. The accuracy and loss curves are then plotted for each model as shown in Figure C-1 to Figure C-5. The mean validation accuracy for each model is shown in Table C-1.

APPENDIX C   SENTIMENT CLASSIFICATION REVISITED

Listing C-1. Sentiment classification using the Twitter US Airline Sentiment dataset

Code:
```
#1. Importing the required libraires
import pandas as pd
import tensorflow as tf
from tensorflow.keras.preprocessing.text import Tokenizer
from tensorflow.keras.preprocessing.sequence import pad_sequences
from tensorflow.keras.models import Sequential
from tensorflow.keras.layers import Embedding, SimpleRNN, Dense, Dropout, GRU, LSTM, Bidirectional
from sklearn.model_selection import train_test_split
from sklearn.preprocessing import LabelEncoder
#2. Load the dataset
data = pd.read_csv("Tweets.csv")
#3. Select relevant columns and drop missing values
data = data[['text', 'airline_sentiment']].dropna()
#4. Encode sentiment labels
label_encoder = LabelEncoder()
data['sentiment'] = label_encoder.fit_transform(data['airline_sentiment'])
#5. Split the data into train and test sets
X_train, X_test, y_train, y_test = train_test_split(data['text'],
data['sentiment'], test_size=0.2)
#6. Tokenize the train and text sequences
max_features = 10000
tokenizer = Tokenizer(num_words=max_features, oov_token='<OOV>')
tokenizer.fit_on_texts(X_train)
X_train_seq = tokenizer.texts_to_sequences(X_train)
X_test_seq = tokenizer.texts_to_sequences(X_test)
#7. Pad the sequences
maxlen = 100
X_train = pad_sequences(X_train_seq, maxlen=maxlen)
X_test = pad_sequences(X_test_seq, maxlen=maxlen)
#8. Model 1
```

```python
model_1 = Sequential([
    Embedding(max_features, 64, input_length=maxlen),
    SimpleRNN(64, return_sequences=False),
    Dense(3, activation='softmax')])
model_1.compile(optimizer='adam', loss='sparse_categorical_crossentropy',
metrics=['accuracy'])
history_1 = model_1.fit(X_train, y_train, epochs=10, batch_size=32,
validation_data=(X_test, y_test))
```
#9. Model 2
```python
model_2 = Sequential([
    Embedding(max_features, 64, input_length=maxlen),
    Bidirectional(SimpleRNN(64, return_sequences=False)),
    Dense(3, activation='softmax')])
model_2.compile(optimizer='adam', loss='sparse_categorical_crossentropy',
metrics=['accuracy'])
history_2 = model_2.fit(X_train, y_train, epochs=10, batch_size=32,
validation_data=(X_test, y_test))
```
#10. Model 3
```python
model_3 = Sequential([
    Embedding(max_features, 64, input_length=maxlen),
    GRU(64, return_sequences=False),
    Dense(3, activation='softmax')])
model_3.compile(optimizer='adam', loss='sparse_categorical_crossentropy',
metrics=['accuracy'])
history_3 = model_3.fit(X_train, y_train, epochs=10, batch_size=32,
validation_data=(X_test, y_test))
```
#11. Model 4
```python
model_4 = Sequential([
    Embedding(max_features, 64, input_length=maxlen),
    LSTM(64, return_sequences=False),
    Dense(3, activation='softmax')])
model_4.compile(optimizer='adam', loss='sparse_categorical_crossentropy',
metrics=['accuracy'])
history_4 = model_4.fit(X_train, y_train, epochs=10, batch_size=32,
validation_data=(X_test, y_test))
```

APPENDIX C   SENTIMENT CLASSIFICATION REVISITED

#12. Model 5
```
model_5 = Sequential([
    Embedding(max_features, 64, input_length=maxlen),
    Bidirectional(LSTM(64, return_sequences=False)),
    Dense(3, activation='softmax')])
model_5.compile(optimizer='adam', loss='sparse_categorical_crossentropy', metrics=['accuracy'])
history_5 = model_5.fit(X_train, y_train, epochs=10, batch_size=32, validation_data=(X_test, y_test))
```
#13. Create a function to plot accuracy and loss curves
```
def plot_history(history, model_name):
    plt.figure(figsize=(12, 6))
    plt.subplot(1, 2, 1)
    plt.plot(history.history['accuracy'])
    plt.plot(history.history['val_accuracy'])
    plt.title(f'{model_name} Model Accuracy')
    plt.xlabel('Epoch')
    plt.ylabel('Accuracy')
    plt.legend(['Train', 'Val'], loc='upper left')
    plt.subplot(1, 2, 2)
    plt.plot(history.history['loss'])
    plt.plot(history.history['val_loss'])
    plt.title(f'{model_name} Model Loss')
    plt.xlabel('Epoch')
    plt.ylabel('Loss')
    plt.legend(['Train', 'Val'], loc='upper left')
    plt.tight_layout()
    plt.show()
```
#14. Plotting accuracy and loss curves for each model
```
plot_history(history_1, "Model 1")
plot_history(history_2, "Model 2")
plot_history(history_3, "Model 3")
plot_history(history_4, "Model 4")
plot_history(history_4, "Model 5")
```
#15. Create a function to calculate mean validation accuracy

APPENDIX C   SENTIMENT CLASSIFICATION REVISITED

```
def mean_validation_accuracy(history):
    val_acc = history.history['val_accuracy']
    mean_acc = np.mean(val_acc)
    return mean_acc
```

#16. Calculate the mean validation accuracy for each model

```
mean_acc_1 = mean_validation_accuracy(history_1)
mean_acc_2 = mean_validation_accuracy(history_2)
mean_acc_3 = mean_validation_accuracy(history_3)
mean_acc_4 = mean_validation_accuracy(history_4)
mean_acc_5 = mean_validation_accuracy(history_5)
```

Output:

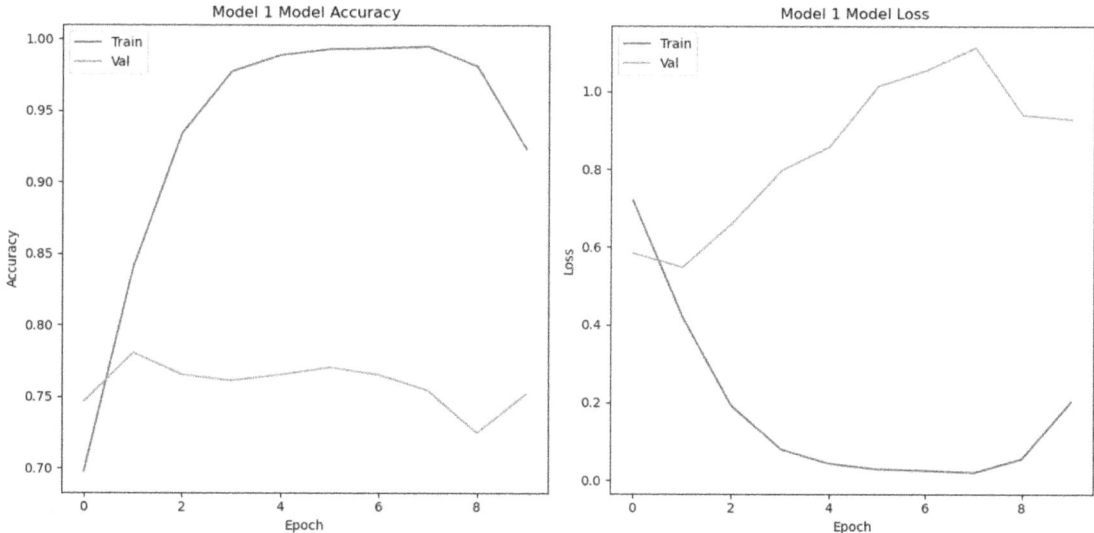

Figure C-1. Loss and accuracy curves: Model 1

APPENDIX C   SENTIMENT CLASSIFICATION REVISITED

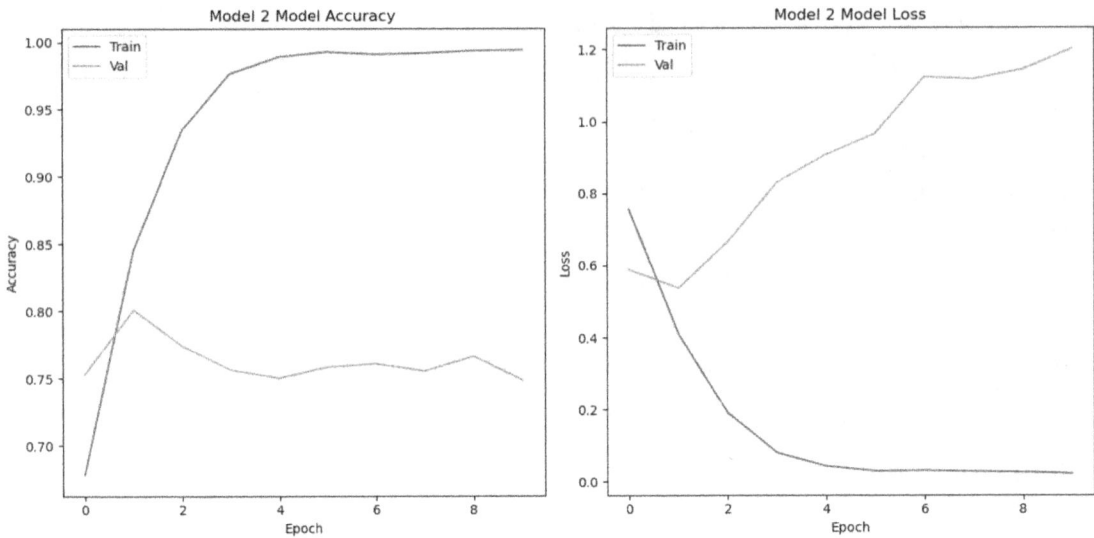

Figure C-2. *Loss and accuracy curves: Model 2*

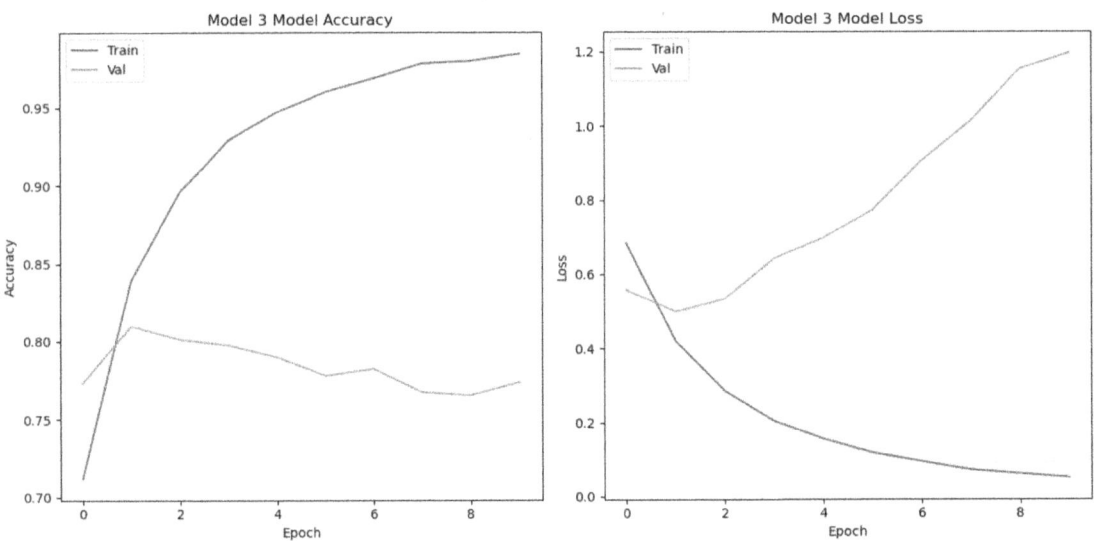

Figure C-3. *Loss and accuracy curves: Model 3*

APPENDIX C   SENTIMENT CLASSIFICATION REVISITED

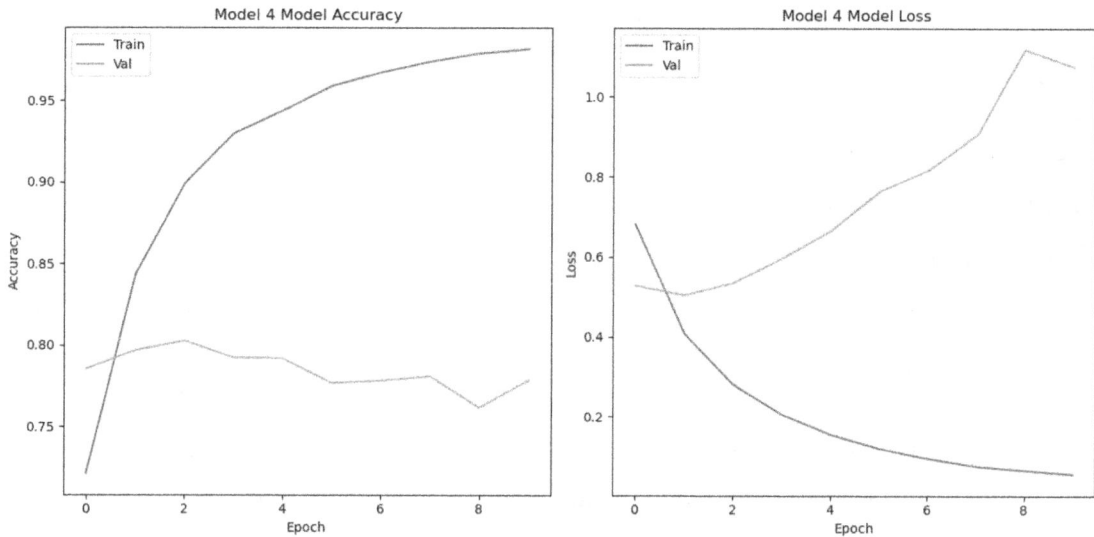

Figure C-4. *Loss and accuracy curves: Model 4*

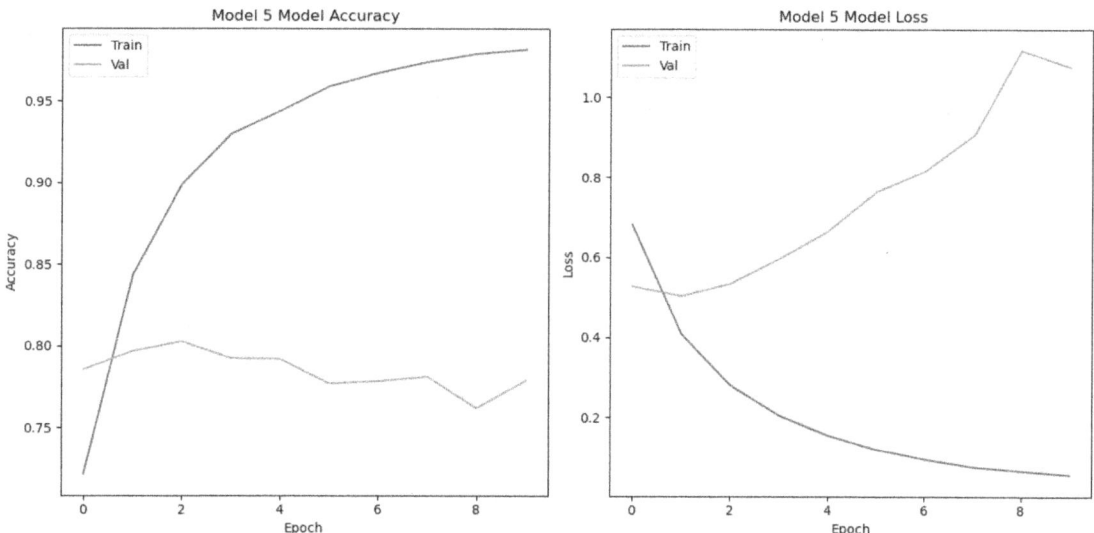

Figure C-5. *Loss and accuracy curves: Model 5*

Table C-1. Mean Validation Accuracy of Five Different Models

Architecture	Mean Validation Accuracy
Simple RNN with a single layer having 64 units	0.7586
Bidirectional RNN with a single layer having 64 units	0.7625
GRU with a single layer having 64 units	0.7839
LSTM with a single layer having 64 units	0.7846
Bidirectional LSTM with a single layer having 64 units	0.7839

The reader is expected to carry out hyperparameter tuning to enhance the performance of the above models and make them more generalizable.

APPENDIX D

Predicting Next Word

We created a text file having around 350 couplets of a famous Urdu poet born around 1797 in Agra, a city in India. We then uploaded the file on the drive. The file had a particular format in which each couplet was followed by two empty lines. The number of characters in the file was counted using the **len** function. This was followed by reading the first few characters (280) of the file:

```
text = open(path_to_file, 'rb').read().decode(encoding='utf-8')
# Number of characters in the file
print('Length of text: ' + str(len(text))+'characters')
print(text[:280])
Output:
ham ko ma.alūm hai jannat kī haqīqat lekin
dil ke khush rakhne ko 'ġhālib' ye khayāl achchhā hai

ishq ne 'ġhālib' nikammā kar diyā
varna ham bhī aadmī the kaam ke

mohabbat meñ nahīñ hai farq jiine aur marne kā
usī ko dekh kar jiite haiñ jis kāfir pe dam nikle
```

We then extracted the unique characters from all these couplets using a set that came out to be 43. This was followed by creating a variable called char_to_id using the String Lookup layer of Keras. The String Lookup layer of Keras converts each character into a particular ID:

```
char_to_id = tf.keras.layers.StringLookup(
    vocabulary=list(vocab), mask_token=None)
id_to_char = tf.keras.layers.StringLookup(
    vocabulary=char_to_id.get_vocabulary(), invert=True, mask_token=None)
```

APPENDIX D   PREDICTING NEXT WORD

Likewise, we created a variable called id_to_char, which converts the ID back to the characters. The so-formed array can be converted to a string using the reduce_join function of strings. Combining these two we created a function called id_to_text, which converts a list of IDs to a corresponding string. We created the database of all the IDs in a variable called id_data:

```
tf.strings.reduce_join(chars, axis=-1).numpy()
```

This was followed by training the RNN in which we kept the sequence length as 128. The batches for training are then created. The input is then split using a function called split_input in which we take the given sequence and extract the target sequence. This function is the same as suggested on the official website of Keras. The function effectively splits the given input into characters. Using the sequence map function, we create the dataset from the above:

```
# Length of the vocabulary in StringLookup Layer
vocab_size = len(char_to_id.get_vocabulary())
# The embedding dimension
embedding_dim = 256
# Number of RNN units
rnn_units = 1024
def __init__(self, vocab_size, embedding_dim, rnn_units):
    super().__init__(self)
    self.embedding = tf.keras.layers.Embedding(vocab_size, embedding_dim)
    self.rnn = tf.keras.layers.SimpleRNN(rnn_units,return_sequences=True,
return_state=True)
    self.dense = tf.keras.layers.Dense(vocab_size)
```

Note that in the input we are giving a small batch, and the target contains the string starting from the second character to the last but one character. We take the batch size of 64 and a buffer size of 1000 to create a dataset as suggested on the official website of Keras.

We then create a class called model_1, which is initialized with the vocabulary size, dimension of embedding, and RNN units. We create an embedding layer, an RNN layer, followed by a dense layer having the same length as the vocabulary size. The summary of the model is as follows:

APPENDIX D PREDICTING NEXT WORD

```
model = Model1(
    vocab_size=vocab_size,
    embedding_dim=embedding_dim,
    rnn_units=rnn_units)
```

The sampled indices are created using the squeeze function of an instance of tf.random.categorical. We use sparse categorical cross-entropy and observe the losses and the mean loss. We compile the model using the Adam optimizer and sparse categorical cross-entropy loss. We run the model through 100 epochs and then run the one step model many times to produce the following output:

Galib:e siyābrā ho saboz o ki rabashā khī daf nahāñ aurñkhar-nabīñ hī jahni chāye-ghzam aatā

ho nahīñ aatā hai-ebānhā gayā tīrān-e-kharat raht raqgh le kaat mujh meññ

hone ke
iThir hoī 'ġhāl-bā-bujā pe hamnijheñ khāhiye
toī kī agakvā
bhīq kahte haiñ tare kahe phin us sī jaamā haq-gufār aur hamāre

se achchhā huā thī kah hattābar nahīñ

aajam ke usm kāsī na sohīnā
vahte gokatī aa.e aur ġharkat ko hat raklā kaheñgeñ mirmat kāte koī

sa kahīñ autchchelit
haz sazār vo chapchhe kiye khamab sahī

dannat-e-ġhaslvauñ hai derab kir nahīñ jar na hāde be-aar hai jaanā par palāhañ iire

idki dish jaanā kahīñ haiñ qiit zānb an dar garā ko barah bahī
nahīñ e-tiyāshā kuchh se jī ballā sahī vakhte hai

ham.asaa hai yahsā kā hāqāte mujhe bait
dormush-e-raa-rakū ros ekrāñkhār-e-lilā.e meñge

APPENDIX D PREDICTING NEXT WORD

```
dil de-hab puchh ue ġhapr-deñ ho thire
hote pai ke abrat kā kī abhīñ k͟hait hī e haftā hai jamān haiñr nahte
haiñ kire
```

Observe the above output. Most of it does not make sense, but have you noticed that it has been able to learn the structure of the poetry? Now develop a next work prediction model and train it on the same dataset. Observe the output. Is it better than the earlier?

Now create a huge dataset of a few thousand couplets and observe if the output has improved.

APPENDIX E

COVID Classification

This appendix presents a CNN model that classifies patients suffering from COVID-19 and healthy controls. The dataset has been obtained from Kaggle (https://www.kaggle.com/datasets/prashant268/chest-xray-covid19-pneumonia) consisting of 1583 images of controls and 576 images of patients. All the images were resized to 224 × 224 to match the input shape. The CNN model contains three convolutional layers each followed by a max pool layer of filter size 2 × 2. This is followed by three dense layers of 128, 64, and 2 (for binary classification) neurons. Listing E-1 implements the above pipeline. The model's loss and accuracy curves are shown in Figure E-1.

Listing E-1. COVID classification using CNN

Code:

#1. Import the required libraries
```
import tensorflow as tf
from tensorflow.keras.applications import VGG19
from tensorflow.keras.models import Model
from tensorflow.keras.layers import Input, Conv2D
from tensorflow.keras.layers import Flatten, Dense, Dropout
from tensorflow.keras.optimizers import Adam
import numpy as np
from sklearn.model_selection import train_test_split
import tensorflow as tf
from tensorflow.keras import datasets, layers, models
import matplotlib.pyplot as plt
```
#2. Load the dataset
```
X = np.load('/content /X.npy')
y = np.load('/content /y.npy')
```

#3. Split the dataset into train and test set

```python
X_train, X_test, y_train, y_test = train_test_split(X, y, test_size = 0.3, shuffle = True)
print(X_train.shape, y_train.shape, X_test.shape, y_test.shape)
```

#4. Create, compile and fit the new model

```python
Model_1 = models.Sequential()
Model_1.add(layers.Conv2D(16, (5,5), activation='relu', input_shape=(224, 224, 1)))
Model_1.add(layers.MaxPooling2D((2, 2)))
Model_1.add(layers.Conv2D(32, (3, 3), activation='relu'))
Model_1.add(layers.MaxPooling2D((2, 2)))
Model_1.add(layers.Conv2D(64, (4, 4), activation='relu'))
Model_1.add(layers.MaxPooling2D((2, 2)))
Model_1.add(layers.Flatten())
Model_1.add(layers.Dense(128, activation='relu'))
Model_1.add(layers.Dense(64, activation='relu'))
Model_1.add(layers.Dense(2, activation='softmax'))
Model_1.compile(optimizer='adam',loss='sparse_categorical_crossentropy',metrics=['accuracy'])
Model_1.summary()
batch_size = 64
history_batch = Model_1.fit(X_train, y_train, epochs=10, batch_size=batch_size, validation_data=(X_test, y_test))
```

#5. Create a function to plot loss and accuracy curve

```python
def plot_history(history, model_name):
    plt.figure(figsize=(12, 6))
    plt.subplot(1, 2, 1)
    plt.plot(history.history['accuracy'])
    plt.plot(history.history['val_accuracy'])
    plt.title(f'{model_name} Model Accuracy')
    plt.xlabel('Epoch')
    plt.ylabel('Accuracy')
    plt.legend(['Train', 'Val'], loc='upper left')
    plt.subplot(1, 2, 2)
    plt.plot(history.history['loss'])
```

```
plt.plot(history.history['val_loss'])
plt.title(f'{model_name} Model Loss')
plt.xlabel('Epoch')
plt.ylabel('Loss')
plt.legend(['Train', 'Val'], loc='upper left')
plt.tight_layout()
plt.show()
```

#6. Plotting accuracy and loss curve for the above model

```
plot_history(history_batch, "CNN")
```

Output:

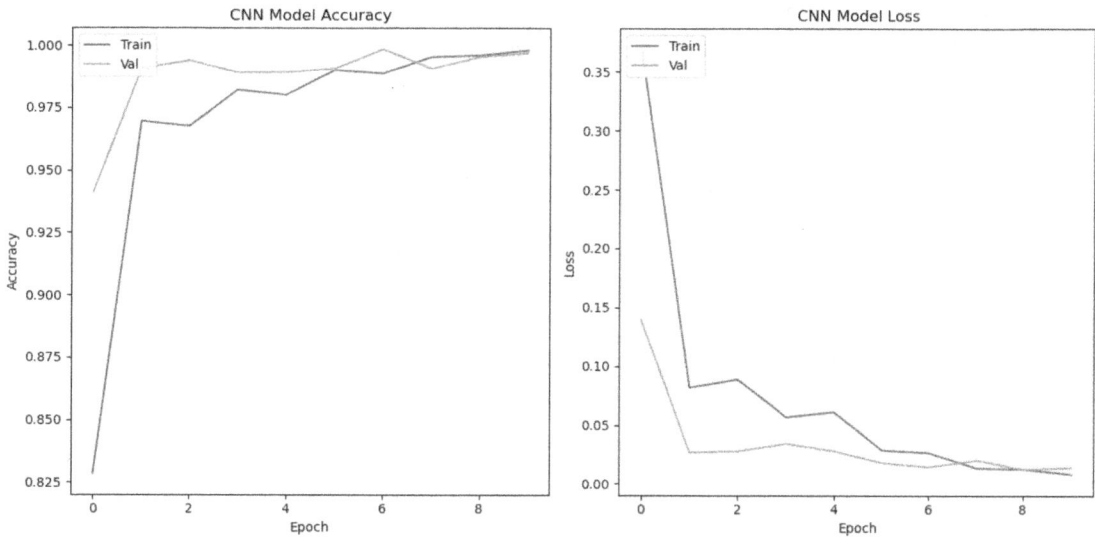

Figure E-1. *Loss and accuracy curves: CNN for COVID classification*

Class Activation Layer

This method helps in finding out the regions of the image responsible for a classification. Assume that you have two sets of images belonging to two different classes. You develop a CNN-based classifier, compile it, train it on the train data, and validate it on the test data. After carrying out hyperparameter tuning, you want to see which region of the image is, on an average, different in the two classes.

APPENDIX E   COVID CLASSIFICATION

Take, for example, a dataset containing chest X-ray images of patients suffering from COVID and controls. You train the model to classify them and then want to see which region of chest X-rays is responsible for this classification so that you can take this image to a radiologist and find out whether bases of classification of your model are good enough. This is a step toward developing an explainable AI model.

In such cases a class activation layer comes to your rescue. This method is based on the heat map representation wherein some pixels are highlighted and associated with a particular class. It uses a global average pooling activation layer, which is placed after the first Convolutional Neural Network layer. This method of finding the discriminating regions is similar to the unsupervised learning model.

The reader is expected to implement the method and find the regions of the X-ray (for the above dataset) responsible for COVID.

Link: `https://www.kaggle.com/code/prameshgautam/class-activation-map-explained`

APPENDIX F

Alzheimer's Classification

This appendix presents a CNN model that classifies patients suffering from Alzheimer's and healthy controls using s-MRI data obtained from OASIS-1. The dataset includes s-MRI scans of 53 controls and 28 patients suffering from Alzheimer's disease. All the images were resized to 224 × 224 to match the input shape. The CNN model contains three convolutional layers each followed by a max pool layer of filter size 2 × 2. This is followed by three dense layers of 128, 64, and 2 (for binary classification) neurons. Listing F-1 implements the above model. The model's loss and accuracy curves are shown in Figure F-1.

Listing F-1. Alzheimer's classification using CNN

Code:
```
#1. Import the required libraries
import tensorflow as tf
from tensorflow.keras.models import Model
from tensorflow.keras.layers import Input, Conv2D
from tensorflow.keras.layers import Flatten, Dense, Dropout
from tensorflow.keras.optimizers import Adam
import numpy as np
from sklearn.model_selection import train_test_split
from tensorflow.keras import datasets, layers, models
import matplotlib.pyplot as plt
#2. Load the dataset
X = np.load('/content /X.npy')
y = np.load('/content /y.npy')
#3. Split the dataset into train and test set
X_train, X_test, y_train, y_test = train_test_split(X, y, test_size = 0.3, shuffle = True)
```

APPENDIX F   ALZHEIMER'S CLASSIFICATION

```python
print(X_train.shape, y_train.shape, X_test.shape, y_test.shape)
```
#4. Create, compile and fit the new model
```python
Model_2 = models.Sequential()
Model_2.add(layers.Conv2D(16, (5,5), activation='relu', input_shape=(224, 224, 1)))
Model_2.add(layers.MaxPooling2D((2, 2)))
Model_2.add(layers.Conv2D(32, (3, 3), activation='relu'))
Model_2.add(layers.MaxPooling2D((2, 2)))
Model_2.add(layers.Conv2D(64, (4, 4), activation='relu'))
Model_2.add(layers.MaxPooling2D((2, 2)))
Model_2.add(layers.Flatten())
Model_2.add(layers.Dense(128, activation='relu'))
Model_2.add(layers.Dense(64, activation='relu'))
Model_2.add(layers.Dense(2, activation='softmax'))
Model_2.compile(optimizer='adam',loss='sparse_categorical_crossentropy',metrics=['accuracy'])
Model_2.summary()
batch_size = 64
history_batch = Model_2.fit(X_train, y_train, epochs=10, batch_size=batch_size, validation_data=(X_test, y_test))
```
#5. Create a function to plot loss and accuracy curve
```python
def plot_history(history, model_name):
    plt.figure(figsize=(12, 6))
    plt.subplot(1, 2, 1)
    plt.plot(history.history['accuracy'])
    plt.plot(history.history['val_accuracy'])
    plt.title(f'{model_name} Model Accuracy')
    plt.xlabel('Epoch')
    plt.ylabel('Accuracy')
    plt.legend(['Train', 'Val'], loc='upper left')
    plt.subplot(1, 2, 2)
    plt.plot(history.history['loss'])
    plt.plot(history.history['val_loss'])
    plt.title(f'{model_name} Model Loss')
    plt.xlabel('Epoch')
```

```
    plt.ylabel('Loss')
    plt.legend(['Train', 'Val'], loc='upper left')
    plt.tight_layout()
    plt.show()
#6. Plot accuracy and loss curve for the above model
plot_history(history_batch, "CNN")
```

Output:

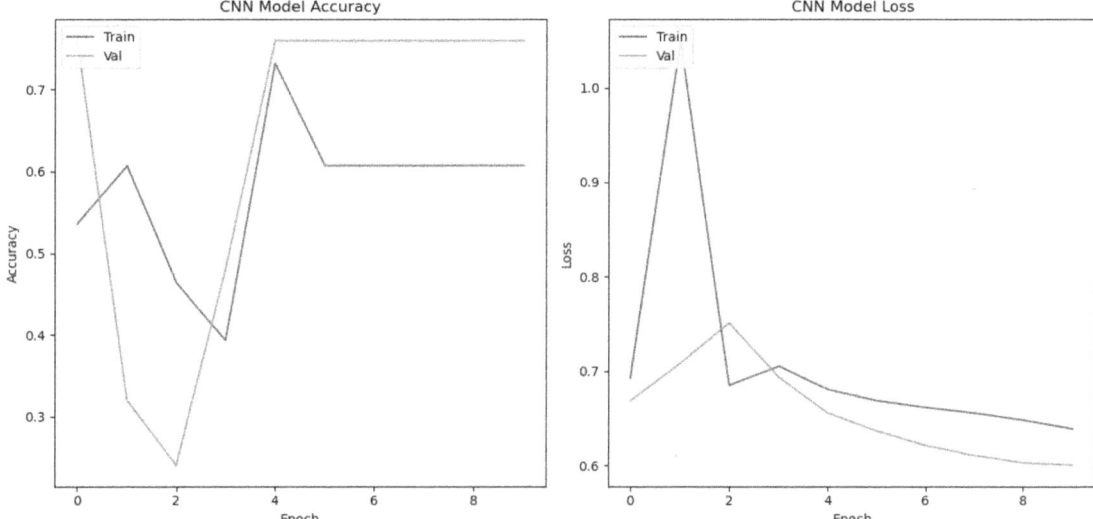

Figure F-1. Loss and accuracy curves: CNN

Note that the above dataset was also classified using transfer learning in Chapter 8. The reader is expected to carry out hyperparameter tuning of the above implementation and find which of the two methods of classification is better in terms of

1. Number of images required for classification
2. Memory required
3. Computation complexity
4. Explainability

APPENDIX G

Music Genre Classification Using MFCC and Convolutional Neural Network

Dataset

The dataset used in this project is George Tzanetakis Music Genre Dataset (GTZAN), obtained from Kaggle. This dataset contains audio files categorized into ten different classes, representing various genres of music, but for this project, we focused on five classes, namely, blues, classical, country, disco, and hip-hop. Each class contains 100 audio samples, making a total of 500 samples. The dataset is split into training and validation sets to implement model training and evaluation.

Feature Extraction

To extract audio features from the dataset, we implement Mel-Frequency Cepstral Coefficients (MFCC).

As per FluCoMa.org,

> *MFCC compresses the overall spectrum into a smaller number of coefficients that, when taken together, describe the general contour of the spectrum.*

This feature extraction method is commonly used in audio processing, which is useful for distinguishing different types of audio signals.

Convolutional Neural Network Architecture

The above step results in images, thus converting the problem into image classification. To classify the audio samples in their respective genres, we implemented a Convolutional Neural Network (CNN) in Listing G-1 with the following architecture:

1. Convolutional Layer (5 × 5):
    - The first layer applies 16 filters with a size of 5 × 5 to the input images.
    - Activation function: ReLU (Rectified Linear Unit).

2. Pooling Layer:
    - A max pooling layer with a size of 2 × 2

3. Convolutional Layer (3 × 3):
    - The second convolutional layer applies 32 filters with a size of 3 × 3.
    - Activation function: ReLU.

4. Pooling Layer:
    - Another max pooling layer with a size of 2 × 2 for further dimensionality reduction

5. Flatten Layer:
    - This layer flattens the 2D matrices into a 1D vector space.

6. Fully Connected Layer:
    - A dense layer with 64 neurons and ReLU activation

7. Output Layer:
    - A dense layer with five neurons for five classes and **softmax** activation for classification

APPENDIX G   MUSIC GENRE CLASSIFICATION USING MFCC AND CONVOLUTIONAL NEURAL NETWORK

The following listing presents a stepwise flow of classifying the above ten classes.

Listing G-1. Music genre classification using MFCC and Convolutional Neural Network

Code:
```
#1. Mount the Google Drive to access the data files
from google.colab import drive
drive.mount('/content/drive')
#2. Import the requisite libraries
import matplotlib.pyplot as plt
import os
import numpy as np
import tensorflow as tf
from tensorflow.keras.models import Sequential
from tensorflow.keras.layers import Conv2D, MaxPooling2D, Flatten, Dense
from sklearn.model_selection import train_test_split
from skimage.transform import resize
#3. Load the dataset stored in Google Drive as Numpy arrays
X = np.load('/content/drive/My Drive/MFCC_Data/X.npy')
y = np.load('/content/drive/My Drive/MFCC_Data/y.npy')
#4.Print and Verify the shape of the data
print('X shape:', X.shape)
print('y shape:', y.shape)
#5. Create a function to resize the images to match the input shape
def resize_images_with_labels(X, y, image_size=(100, 400)):
    resized_images = []
    for img in X:
        # Resize the image
        resized_img = resize(img, (image_size[0], image_size[1], 4), anti_
        aliasing=True)
        resized_images.append(resized_img)
    resized_images_array = np.array(resized_images)
    labels_array = np.array(y)
    return resized_images_array, labels_array
images_array, labels_array = resize_images_with_labels(X, y)
```

APPENDIX G MUSIC GENRE CLASSIFICATION USING MFCC AND CONVOLUTIONAL NEURAL NETWORK

```
images_array = images_array[:, :, :, :3]
print(images_array.shape)
print(labels_array.shape)
```
#6. Create CNN Model
```
model = Sequential([Conv2D(16, (5, 5), activation='relu', input_shape=(X.shape[1], X.shape[2], X.shape[3])),
    MaxPooling2D((2, 2)),
    Conv2D(32, (3, 3), activation='relu'),
    MaxPooling2D((2, 2)),
    Flatten(),
    Dense(64, activation='relu'),
    Dense(5, activation='softmax')
])
model.compile(optimizer='adam', loss='sparse_categorical_crossentropy', metrics=['accuracy'])
model.summary()
```
#7. Split the dataset into train and test set
```
X_train, X_val, y_train, y_val = train_test_split(X, y, test_size=0.2, random_state=42)
```
#8. Fit the compiled model on the train set
```
m1=model.fit(X_train, y_train, epochs=10, batch_size=16, validation_data=(X_val, y_val))
```

The reader is expected to analyze the loss and performance curves and explore the possibilities of improving the performance of the model. It may be noted that the model results in an accuracy of 0.95 with the current dataset.

Index

A

AD, *see* Alzheimer's disease (AD)
Adam optimizer, 118, 119
ADNI, *see* Alzheimer's Disease
	Neuroimaging Initiative (ADNI)
AI, *see* Artificial Intelligence (AI)
AlexNet, 157, 185
	code, 199, 200
	features, 198
	ImageNet, 199
	overfitting, 200
	structure, 199, 200
Alzheimer's disease (AD), 214
Alzheimer's Disease Neuroimaging
	Initiative (ADNI), 207
Area under the Receiving Curve (AUC), 9
Artificial Intelligence (AI), 4
AUC, *see* Area under the Receiving
	Curve (AUC)
Autoencoder
	exercises, 304–306
	experiments, 293–295, 297–299
	implementation, 287
	math, 288
	PCA, 290, 291
	representation, multiple layers, 300, 301
	training, 291, 292
	types
		over-complete, 289, 290
		user-complete, 289
	variants
		denoising, 303
		hidden layer, 303
		sparse, 302
		variational, 303

B

Backpropagation algorithm, 59, 86, 104, 111, 192
Backpropagation Through Time (BPTT), 228, 284
Batch gradient descent (BGD), 114, 127
BGD, *see* Batch gradient descent (BGD)
Bias, 137
BMs, *see* Boltzmann Machines (BMs)
Boltzmann Machines (BMs), 318, 319
BPTT, *see* Backpropagation Through
	Time (BPTT)

C

CD, *see* Contrastive Divergence (CD)
ChatGPT, 314, 318
Class activation layer, 349, 350
CNN, *see* Convolutional Neural
	Network (CNN)
Contrastive Divergence (CD), 314
Convolutional neural networks (CNNs), 47, 133, 257, 323
	Alzheimer disease, 351, 353
	architecture, 356–358
	components, 158, 159
	convolutional layer, 159–161, 163, 165
	definition, 157

INDEX

Convolutional neural networks
  (CNNs) (*cont.*)
  exercises, 182–184
  fully connected layer, 170
  hyperparameters, 157
  kernels, 170–172, 174, 175, 177
  LeNet, 177–181
  MNIST dataset, 192
  neurocognition, 181
  normalization, 169
  padding, 165–167
  pooling layer, 168, 169, 181
  sequential model, 185
  stride, 167
COVID classification
  CNN, 347, 348
  loss and accuracy, 349

D

Deep learning
  AI, 45, 46
  exercises, 56, 57
  generate data, 52
  imagery/convolutional neural
    network, 47–49
  neurons, 43–45
  optimization algorithms, 50
  representation-learning methods, 52
  sequences, 50
Deep Neural Networks (DNNs), 45, 133
Denoising autoencoders, 303
Dense Neural Networks (DNNs), 257
DNNs, *see* Deep Neural Networks (DNNs)

E

Exploding gradient, 78

F

Face detection
  find_face, 332, 333
  MTCNN, 331
FDR, *see* Fisher Discriminant Ratio (FDR)
Fisher Discriminant Ratio (FDR), 14

G

Gated Recurrent Unit (GRU), 51, 251
  architecture, 259
  vanishing gradient, 258
Gemini, 55
Generative models
  boltzmann machines, 310–314
  exercises, 318–321
  Hopfield networks, 307–310
  supervised and unsupervised
    learning, 307
  transformers, 314–317
Genetic algorithms, 14
George Tzanetakis Music Genre Dataset
  (GTZAN), 355
GLCM, *see* Gray-Level Co-occurrence
  Matrix (GLCM)
Google LeNet, 185
Google maps, 5
GoogLeNet, 207
  DenseNet, 202
  inception module, 201
  ResNet, 201
  RmsProp optimizer, 201
Gray_image, 21
Gray-Level Co-occurrence Matrix
  (GLCM), 20
GRU, *see* Gated Recurrent Unit (GRU)
GTZAN, *see* George Tzanetakis Music
  Genre Dataset (GTZAN)

H

Handwritten digit classification, 38
Handwritten text recognition, 234, 249
Heuristic search algorithms, 14
Histogram of Oriented Gradients, 23
Hyperparameter tuning
    autoencoders, 141
    bias-variance, 134–137
    CNN, 140, 141
    definition, 133
    DNN, 137, 140
    exercises, 150–154
    experiments, 142–145, 147, 149, 150
    sequence models, 141
    training data, 150

I

Image captioning, 234
ImageNet, 48, 198, 208
Inception V1, 201

J

Jordan network, 50

K

Kaggle, 335
Keras, 181, 185, 202, 331
    activations, 190, 191
    Conv2D, 190
    dense, 189
    initializing weights, 191
    pooling, 190
keras.Sequential method, 186
Kernels, 170
K-fold splitting technique, 113

K-Nearest Neighbors (KNN), 31
KNN, *see* K-Nearest Neighbors (KNN)

L

Large Language Models (LLMs), 55, 314
layers.add method, 187
layers.pop method, 187
LBP, *see* Local Binary Pattern (LBP)
LeNet, 47, 157, 177
    backpropagation, 192
    implementation, 194–198
    structure, 192–194
Linear regression, 12
LLMs, *see* Large Language
    Models (LLMs)
Local Binary Pattern (LBP), 3, 21
Long Short-Term Memory (LSTM), 51,
    251, 258, 260, 261
LSTM, *see* Long Short-Term
    Memory (LSTM)

M

Machine Learning (ML)
    applications, 5
    bias-variance trade-off
        bias/variance, 29, 30
        overfitting/underfitting, 28
        parameter, 28
    definition, 4
    exercises, 39–41
    feature extraction
        GLCM, 20
        LBP, 21–23
        oriented gradients, histogram, 24
        text data, 19
        types of features, 20

INDEX

Machine Learning (ML) (*cont.*)
    feature selection methods, 14
        filter, 14–17
        filter *vs.* wrapper, 19
        wrapper, 18
    handwritten digits, 31–33, 35, 36, 38
    history, 3
    MNIST dataset, 2, 3
    performance, 7, 9–11
    performance measure, 4
    pipeline, conventional, 11, 12
    pixels, 1
    principal component analysis, 24–27
    regression, 12, 13
    types, 6, 7
Machine translation, 234
Matplotlib, 331
McCulloch–Pitts model, 55
Mel-Frequency Cepstral Coefficients (MFCC), 355
MFCC, *see* Mel-Frequency Cepstral Coefficients (MFCC)
Mini-batch gradient descent, 114
ML, *see* Machine learning (ML)
MLP, *see* Multi-layer perceptron (MLP)
Multi-layer perceptron (MLP), 46, 157
    architecture, 82–84
    backpropagation, 86, 87
    gradient descent, 84, 85
    implementation, 87–94, 96, 97, 99, 101, 102, 104
    XOR problem, 80, 81

N

Named Entity Recognition (NER)
    code, 262, 263, 265–268
    CoNLL-2003 dataset, 262
    loss and accuracy, 268–271
    mean validation accuracy, 272
    sentiment classification, 273, 275–282
    softmax activation, 262
NER, *see* Named Entity Recognition (NER)
Neural networks, 31
    activation functions
        ReLU, 78
        sigmoid, 76
        softmax, 79
        tanh, 77
    exercises, 105–108
    implementation, SLP, 64, 65, 67, 68, 70–74
    neuron structure, 59–61
    numerical, 109
    SLP, 62, 63
    XOR problem, 75
Nonlinear regression, 134

O

Overfitting, 136
Overlapping window, 227

P, Q

Parts of speech (POS) tagging, 241
PCA, *see* Principal Component Analysis (PCA)
Perceptrons, 45
Predicting next word
    char_to_id, 343
    id_to_char, 344
    len function, 343
    sequence map function, 344
    squeeze function, 345
Principal Component Analysis (PCA), 3, 290

R

RBMs, *see* Restricted Boltzmann Machines (RBMs)
Recurrent Neural Network (RNN), 257, 285, 314, 335
    applications
        handwritten text recognition, 249
        POS tagging, 241, 243–246, 248
        sentiment classification, 234–236, 238–240
        speech to text, 250, 251
    BPTT, 229
    exercises, 251–254
    neural network, sequences, 226, 227
    sequence data, 226
    time intervals, sequence depicting, 225
    time stamps, 228
    types, 230–233
Recursive Feature Elimination (RFE), 18
Recursive network, 258
reduce_join function, 344
Regression, 12
Reinforcement learning, 7
ReLU activation, 191, 198
ResNet, 201
Restricted Boltzmann Machines (RBMs), 318, 319
RFE, *see* Recursive Feature Elimination (RFE)
RMSprop, 111, 120, 125, 126
RNNs, *see* Recurrent Neural Networks (RNNs)
Rosenblatt Perceptron model, 61

S

Semi-supervised learning (SSL), 7
Sentiment analysis, 234
Sentiment classification
    hyperparameter tuning, 342
    Kaggle, 335
    Twitter US Airline sentiment dataset, 336, 338–341
Sequential model, CNN
    adding layers, 187
    creating model, 186, 187
    initializing weights, 188
    removing layers, 187
    TITO, 186
SGD, *see* Stochastic gradient descent (SGD)
Sigmoid activation function, 76
Simpsons characters
    Adam optimizer, 325
    CNN, 323, 324
    loss curve, 326, 327, 329
Single-Layer Perceptron (SLP), 61, 62
sklearn.neural_network.MLPClassifier function, 87
SLP, *see* Single-Layer Perceptron (SLP)
Softmax activation function, 191
Sparse autoencoder, 302
Speech-to-text conversion, 234
SSL, *see* Semi-supervised learning (SSL)
Stochastic gradient descent (SGD), 68, 114, 127
Stride, 160
Supervised learning, 6
Support Vector Machine (SVM), 3, 31
SVM, *see* Support Vector Machine (SVM)

T

Tanh activation function, 77
Tensor Input Tensor Output (TITO), 186, 189, 203

INDEX

Test data, 111
TITO, *see* Tensor Input Tensor Output (TITO)
Training data, 111
Training deep networks
    Adam optimizer, 118–122, 125, 126
    BGD, 114
    exercises, 127–130
    k-fold split, 112, 113
    mini-batch gradient descent, 114–116
    RMSprop, 117
    stochastic gradient descent, 114
    train-test split, 111
    train-validation, 112
Transfer learning, 52
    exercises, 220–222
    limitations/applications, 219
    types/strategies, 217–219

VGG 16, 208, 209, 212, 213
VGG 19, 208, 210–212, 214, 216

U

Under-complete autoencoders, 289
Underfitting, 136
Unsupervised learning, 6

V

Validation set, 112
VAE, *see* Variational autoencoder (VAE)
Variational autoencoder (VAE), 303
VGG 16 model, 208

W, X, Y, Z

Wrapper methods, 18

GPSR Compliance
The European Union's (EU) General Product Safety Regulation (GPSR) is a set of rules that requires consumer products to be safe and our obligations to ensure this.

If you have any concerns about our products, you can contact us on

ProductSafety@springernature.com

In case Publisher is established outside the EU, the EU authorized representative is:

Springer Nature Customer Service Center GmbH
Europaplatz 3
69115 Heidelberg, Germany